John Joseph Griffin

The Life and Times
of an Irish Emigrant

Hugh George Griffin

Copyright © 2012
Savanna Press, U.K.

www.savannapress.com

British Library Cataloguing-in-Publication Data

A catalogue record for this book is available from the British Library

ISBN: 978-0-9571229-0-1

We very much welcome feedback from our readers. Please contact us
if you notice any errors or can provide any new or updated information
relating to any topic covered by this book. Your suggestions will help
us compile a new edition. Please contact us at: www.savannapress.com

Cover picture: Signpost pointing to Maroonah Station in Western
Australia. Photograph © Yoji Murakawa.

Contents

Pictures and Records

Abbreviations

AIF	Australian Imperial Force
ANZAC	Australian and New Zealand Army Corps
c.	circa, approximately (referring to a date)
CE	Church of England
Co.	County
COI	Church of Ireland
EPIP	Eight person Indian production (a military tent)
GS	Gunshot
GSW	Gunshot wound
GW	Gunshot wound
HMAT	His Majesty's Australian Transport (referring to a ship)
HS	Hospital ship
KIA	Killed in action
MIA	Missing in action
NOK	Next of kin
N/R	Not recorded
Pte.	Private (referring to a soldier)
RC	Roman Catholic
[*sic*]	Intentionally so written (used to indicate a potential error in the original source)
SS	Steamship
SW	Shell wound (referring to a battlefield injury)
WA	Western Australia
WIA	Wounded in action

Glossary of Terms

ANZAC day: 25th April. A national day of remembrance in Australia and New Zealand. The acronym ANZAC refers to the Australian and New Zealand Army Corps.

Bioscope: A fairground or music hall attraction consisting of a travelling cinema. The heyday of the bioscope was from the late 1890s until World War I.

Boer War: A war in southern Africa between the Afrikaans-speaking Dutch settlers and the forces of the British Empire, 1899-1902.

Drover: A person, usually an experienced stockman, who moves livestock "on the hoof" over long distances.

Flying doctor: An emergency and primary health care service for those living in remote areas of Australia. The service began in 1928.

Grazier: The owner of a large agricultural property known as a station.

Jackaroo: A trainee stockman.

Ladies companion: A woman from a respectable family who acted as a paid companion for women of rank or wealth.

Limelight: A type of stage lighting once used in theatres and music halls.

Locks: After shearing, the wool is separated into categories: fleece (which makes up the vast bulk), pieces, bellies, locks and crutchings.

Magic lantern: An early type of image projector developed in the 17th century.

Pastoral lease: An Australian land apportionment system created in the mid-19th century to facilitate the division and sale of land to indigenous Australians and European colonists.

Pastoralist: The owner of a large agricultural property known as a station.

Service record: Documentation of the activities and accomplishments of a member of the armed forces.

Squatter: In Australia the term referred to a landowner who grazed livestock on a large scale (whether the station was held by leasehold or freehold title).

Station: A large Australian landholding used for livestock production, often thousands of square miles in area, with the nearest neighbour being hundreds of miles away.

Station hand: An employee involved in routine duties on a station including caring for livestock.

Station overseer: Manager of a station.

Stockman: A person who looks after the livestock on a station. A trainee stockman is known as a jackaroo (male) or jillaroo (female).

Timeline

1830	
	1834 Birth of Edward Griffin (father of John Joseph Griffin), Cork, Ireland
1840	
	1844 Birth of John Harman Mansfield (uncle of John Joseph Griffin)
	1845 - 1849 **Irish Potato Famine**
1850	
	1856 Birth of Elizabeth Mansfield (mother of John Joseph Griffin), Kerry, Ireland
1860	
1870	
	1872 John Harman Mansfield emigrates to Australia
1880	1880 Marriage of Edward Griffin and Elizabeth Mansfield, Killorglin, Ireland
	1881 Birth of John Joseph "Jack" Griffin, Killorglin, Ireland
	1882 Birth of Margaret Catherine "Madge" Griffin, Killorglin, Ireland
	1885 Birth of George Griffin, Killorglin, Ireland
	1886 Birth of Edward Blake "Ned" Griffin, Killorglin, Ireland
1890	
	1899 - 1902 **Boer War**
1900	1900 Jack emigrates to Australia arriving 16th February 1901
	1903 Marriage of Margaret Catherine Griffin and Joseph Naughtin [*sic*], Killorglin, Ireland
	1907 Purchase of Maroonah station by John Joseph Griffin
	1909 Marriage of John Joseph Griffin and Mary Glass
1910	1910 Death of Mary Griffin *née* Glass
	1913 Repossession of Maroonah Station
	1913 Marriage of Margaret Catherine Naughton *née* Griffin and Edward Jones
	1914 - 1918 **World War I**
	1915 John Joseph Griffin enlists in AIF
	1916 John Joseph Griffin wounded in action (Somme Offensive)
	1917 John Joseph Griffin discharged from AIF on medical grounds
1920	
	1926 Death of John Joseph Griffin, Roebourne, Western Australia
1930	

Acknowledgements

I would like to express my gratitude to relatives and family members for providing material that I have included in this book. In particular I would like to thank Claire Cornmell, Stephen Jones and Shaun Glass for providing various documents and photographs and for their inspiration, support and encouragement.

I would also like to thank the following: Peter Dennis, Emeritus Professor of the AIF Project, for permission to use the AIF database entry on John Joseph Griffin; Gregory Cope, Copyright Officer of the National Archives of Australia for permission to reproduce material from the service file, B2455, of John Joseph Griffin; Francine Bull of *The West Australian* for permission to reproduce articles from the newspaper; Paul Johnson of the National Archives Image Library Manager (UK) for permission to publish ships' passenger lists; Aideen M. Ireland of the Reader Services Division, National Archives of Ireland for granting permission to quote from records in the custody of the National Archives of Ireland including the 1901 and 1911 census information; David Biggins of angloboer.com for permission to use material from the angloboer website; Liam Kelly of Findmypast for permission to reproduce images; Stephen Tully for the photos of the Hawthorn Leslie shipyard; Rob Nelson, Perth WA, for permission to reproduce material from *Migrant Ships Arriving In Queensland 1837-1915*; Ian Cruttenden, for permission to include quotations from the war diary of Herbert Charles Cruttenden; and Yoji Murakawa for providing the photograph on the front cover. Irish Genealogy provided many of the church records of Irish baptisms, marriages and deaths. The photograph of the Queensland Mounted Infantry was provided by the John Oxley Library, State Library of Queensland. Thanks also to Loreley A. Morling, Genealogical and Historical Researcher, for performing research on my behalf at the State Records Office in Perth, Western Australia and to Bev Russell, Research Officer at Western Australian Genealogical Society for help and advice.

Most of all, I would like to thank my wife Annette for her personal support and great patience at all times.

Hugh Griffin

Chapter 1: Ireland

Kerry in 19th Century

John Joseph Griffin, known to his family as Jack, was born on the 9th of February 1881 in Co. Kerry in the south-west of Ireland. His father had been born in neighbouring Co. Cork in 1834. In 1845, when Jack's father was eleven years old, the Irish potato crop failed. At this time, Kerry, in common with most of rural Ireland, was populated by poor tenant farmers who relied on the potato as their main food. During the Great Irish Famine of 1845 to 1849 large numbers of people emigrated to seek a better life overseas. Many who remained in Ireland died of starvation and malnutrition. Some estimates put the number of deaths attributed to the famine as high as one million[1]. The famine led to the Land War of the 1870s and 1880s, in which tenant farmers fought for better terms from their landlords.

At the time of Jack's birth in 1881, rural areas of Co. Kerry were experiencing endemic poverty and people suffered a subsistence living which occasionally came close to famine[2]. Generations of people were forced to leave in search of a basic living. Around the year 1872 Jack's uncle, John Harman Mansfield, emigrated from his home in Co. Kerry to Western Australia where he lived until his death in 1907.

Kerry had the highest annual rate of emigration from Ireland[2]. Census data shows that the population of the county was 295,000 in 1841 and this had fallen to 160,000 by 1911. The population continued to fall for much of the twentieth century. People emigrated to England, America, Australia and many other countries. Today, the population of Kerry is only 145,048 (2011 census), half the population of 1841.

The departure points for most Kerry people were the various railway stations located throughout the county. In 1853, Killarney

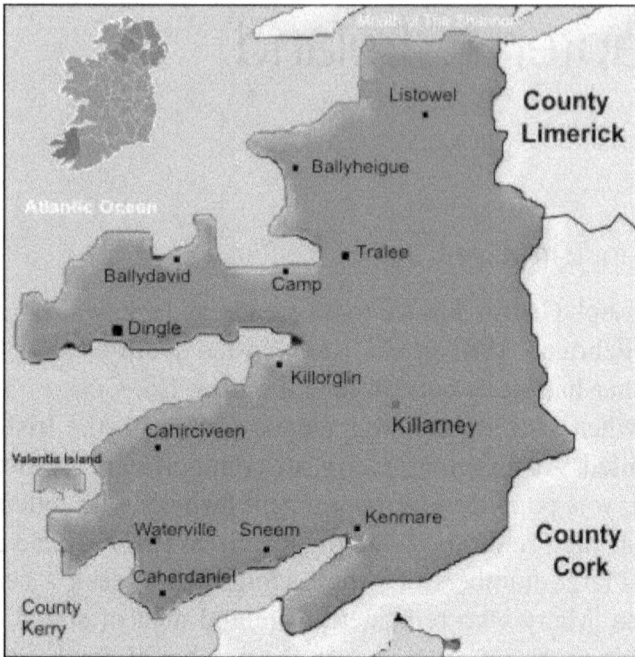

Co. Kerry, Ireland

was connected to the main Dublin-Cork line at Mallow, and by 1911 almost the entire county had rail services.

At this time, Roman Catholics made up more than 95 per cent of the population of Co. Kerry[2]. There were just 3,623 Church of Ireland (Anglican) members, 251 Methodists, 249 Presbyterians, 26 Jews, 67 members of various other assorted religions, and two people who refused to disclose what, if any, religion they held[2].

The Griffins of Killorglin

Jack's father, Edward Griffin, was a school teacher who came from County Cork, Ireland. His mother, Elizabeth (*née* Mansfield) was from Kerry, the daughter of Joseph Mansfield and Margaret Harman. The couple married on the 18th of April 1880 in the Church of Ireland parish of Killorglin, Co. Kerry (see marriage record on page 3).

Killorglin c. 1900. This photograph was taken during Puck Fair, an annual street festival held in Killorglin.

Area - KERRY (COI) ,
Parish/Church/Congregation - KILLORGLIN

Marriage of EDWARD GRIFFIN of KILLORGLIN and ELIZABETH MANSFIELD of KILLORGLIN RECTORY on 18 April 1880

	Husband	Wife
Name	EDWARD GRIFFIN	ELIZABETH MANSFIELD
Address	KILLORGLIN	KILLORGLIN RECTORY
Occupation	TEACHER	TEACHER
Father	JOHN GRIFFIN	JOSEPH MANSFIELD
Mother	N/R	N/R

Further details in the record

Recorded Diocesan Area	ARDFERT & AGHADOE
Recorded Parochial Area	KILLORGLIN
Priest	NR
Husband's Father's Occupation	TRADESMAN
Wife's Father's Occupation	TEACHER
Witness 1	DORA EAGAR
Witness 2	REBECCA MANSFIELD

Record of the marriage between Edward Griffin and Elizabeth Mansfield. **From:** *Irish Genealogy.*

John Griffin —— Margaret Blake

Edward —— Elizabeth Mansfield

Mary Glass* —— John Joseph (Jack)*

George —— Edward Blake —— Mary Jane Agnew

Leila | George Mansfield | Geraldine (Dena) | Mary (May) | Dorothy (Dorrie) | Maude | Louisa (Louie) | Hilda | **Edward (Ted)***

Edward (Ted) Jones*

Margaret Catherine (Madge)*

Joseph Edward Naughton

Thomas Edward*

Hilda Constance | Mavis Kathleen | Alwyn Edward (Ted)

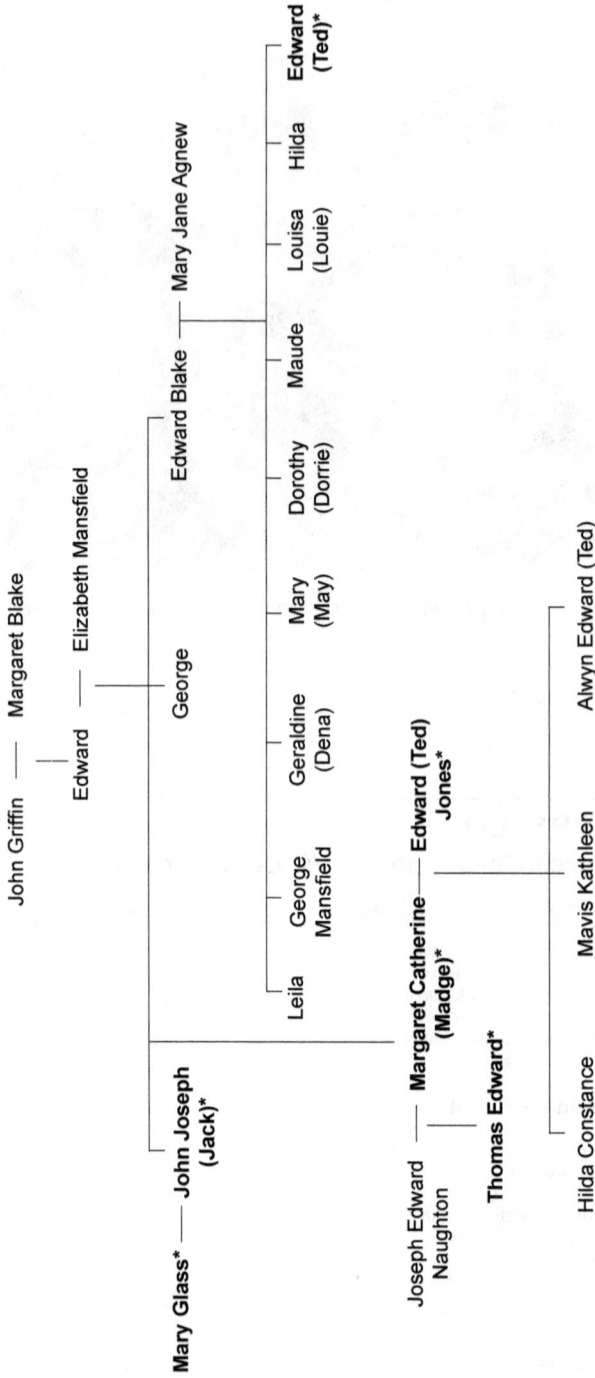

* emigrated to Australia

The Griffins. **Bold type** *indicates the individuals from this family that emigrated to Australia.*

Jack's mother, Elizabeth Griffin née Mansfield.

> **Area - CORK & ROSS (RC) ,**
> **Parish/Church/Congregation - SKIBBEREEN**
> **(CREAGH & SULLON)**
>
> Baptism of EDWARD GRIFFIN of N/R on 29 July 1834
>   **Name**   EDWARD GRIFFIN
> **Date of Birth** N/R N/R N/R
>   **Address**   N/R
>   **Father**   JN GRIFFIN
>   **Mother**   MGT BLAKE
>
> **Further details in the record**
>
>   **Sponsor 1**   JOS WM GRIFFIN
>   **Sponsor 2**   SARAH SHAND
> **Recorded Parochial Area** SKIBBEREEN (CREAGH AND SULLON)

*Baptism record (RC) of Edward Griffin. **From:** Irish Genealogy.*

Edward had been baptised a Roman Catholic in the parish of Skibbereen (Creagh and Sullon), Co. Cork, on the 29th of July, 1834 (see baptism document, above). His marriage, however, was a Church of Ireland affair. Family recollections[3] recount that Edward Griffin had commenced training to become a catholic priest but became disillusioned with Catholicism and, at the age of 45, married his protestant bride.

Their first child, John Joseph ("Jack"), was born on the 9th of February 1881. Jack was baptised in the Church of Ireland congregation of Killorglin on the 13th March 1881 (see baptism

> **Area - KERRY (COI) ,**
> **Parish/Church/Congregation - KILLORGLIN**
>
> Baptism of JOHN JOSEPH GRIFFIN of PAROCHIAL SCHOOLHOUSE on 13 March 1881
>   **Name**   JOHN JOSEPH GRIFFIN
> **Date of Birth** 9 February 1881
>   **Address**   PAROCHIAL SCHOOLHOUSE
>   **Father**   EDWARD GRIFFIN
>   **Mother**   ELIZABETH NR
>
> **Further details in the record**
>
> **Father Occupation** SCHOOL MASTER

*Baptism record of John Joseph "Jack" Griffin. **From:** Irish Genealogy.*

Edward Griffin ——— Elizabeth Mansfield

— John Joseph "Jack"
— Margaret Catherine "Madge"
— George (died in infancy)
— Edward Blake "Ned"

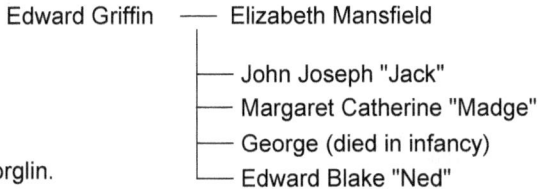

The Griffins of Killorglin.

record, page 6). His address is given as the parochial schoolhouse. As his father, Edward, was a schoolmaster presumably the family lived in accommodation attached to the school.

Jack's sister, Margaret Catherine ("Madge") was born on the 2nd of July the following year. She married Joseph Edward

Jack's sister, Margaret Catherine Griffin (Madge), age 17, Tralee, Ireland.

**Area - KERRY (COI) ,
Parish/Church/Congregation - KILLORGLIN**

Baptism of MARGRET C. GRIFFIN of PAROCHIAL SCHOOLHOUSE on 16 July 1882

    Name   MARGRET C. GRIFFIN

Date of Birth 2 July 1882

    Address   PAROCHIAL SCHOOLHOUSE

    Father   EDWARD GRIFFIN

    Mother   ELIZABETH NR

Further details in the record

Father Occupation SCHOOL MASTER

*Baptism record of Margaret Catherine "Madge" Griffin. **From:** Irish Genealogy.*

**Area - KERRY (COI) ,
Parish/Church/Congregation - KILLORGLIN**

Baptism of GEORGE GRIFFIN of PAROCHIAL SCHOOLHOUSE on 16 November 1885

    Name   GEORGE GRIFFIN

Date of Birth 13 November 1885

    Address   PAROCHIAL SCHOOLHOUSE

    Father   EDWARD GRIFFIN

    Mother   ELIZABETH NR

Further details in the record

Father Occupation SCHOOLMASTER

*Baptism record of George Griffin. **From:** Irish Genealogy.*

**Area - KERRY (COI) ,
Parish/Church/Congregation - KILCOLMAN**

Burial of GEORGE GRIFFIN of KILLORGLIN on 17 November 1885

    Name   GEORGE GRIFFIN

    Address   KILLORGLIN

    Age   NR

Date of Death 17 November 1885 (BASED ON OTHER DATE INFORMATION)

    Occupation NR

Further details in the record

Denomination C. OF I.

*Record of the death of George Griffin. **From:** Irish Genealogy.*

Area - KERRY (COI) ,
Parish/Church/Congregation - KILLORGLIN

Baptism of EDWARD BLAKE GRIFFIN of PAROCHIAL SCHOOLHOUSE on 1 January 1887

Name   EDWARD BLAKE GRIFFIN
Date of Birth 1 December 1886
Address   PAROCHIAL SCHOOLHOUSE
Father   EDWARD GRIFFIN
Mother   ELIZABETH NR

Further details in the record

Father Occupation SCHOOLMASTER

---

*Baptism record of Edward Blake "Ned" Griffin. **From:** Irish Genealogy.*

Naughton and they had a son, Thomas Edward Naughton. Madge's husband, Joseph, died several years later and Madge emigrated to Australia with her young son.

Jack's brother George was born on the 13th November 1885. He was baptised on the 16th November and, sadly, died the following day.

From: AIF Service Record. Jack specifies the three years spent working in R. Hilliard, Ironmongers, New St., Killarney, Ireland.

Joseph Mansfield —— Margaret Harman

Annie — **John Harman***   Elizabeth — Edward   Rebecca — Richard
McCafferty*   Griffin   Hilliard

Daniel Dixon — Margaret   Catherine — James Grey

George Harman*   **Anne***

John Joseph*   George   Edward Blake

Margaret Catherine* — Joseph Naughton

Thomas*

* emigrated to Australia

The Mansfields. **Bold type** *indicates the individuals from this family that emigrated to Australia. Dotted vertical lines indicate siblings that do not feature in this narrative (further details can be found in Jones, A.E. 1998 "Mansfield of Maroonah" ISBN 0957788355).*

Residents of a house 16 in New Street (Lower)

Surname	Forename	Age	Sex	Relation to head	Religion	Birthplace	Occupation
Symons	Louise K	24	Female	Governess	Church of Ireland	Co Sligo	Governess
Hilleard	Richard	50	Male	Head of Family	Church of Ireland	Killarney, Co Kerry	Ironmonger
Barry	Margaret	20	Female	Servant	Roman Catholic	Co Kerry	General Servant
Naughton	John	20	Male	Shop Man	Church of Ireland	Co Kerry	Shop Man
Hilleard	R Franklen	12	Male	Son	Church of Ireland	Co Kerry	Scholar

1901 census data for 16 New Street Killarney. Richard Hilleard [sic] is shown as head of family and his occupation is given as ironmonger. Also at this address is a John Naughton, possibly a younger brother of the Joseph Edward Naughton who married Madge Griffin in 1903. By the time this census was taken, Jack had left Ireland and was living in Western Australia. From: http://www.census. nationalarchives.ie

Jack's youngest brother, Edward Blake ("Ned"), was born on the 1st December 1886. Ned married Mary Jane Agnew and they had nine children. Their descendants live in Ireland, the UK, Africa and Australia.

As his father was a school master, we can assume that Jack Griffin received at least an adequate education in Killorglin. As a young man he spent several years working for R. Hilliard, an ironmonger in New Street, Killarney, Co. Kerry (see page 9). Killarney is about fifteen miles from his native Killorglin so it is likely that, during this time, he lived in Killarney. The 1901 census lists Richard Hilliard and his family (Church of Ireland) living at 16 New Street (see page 10) and a 1913 phone book for the area lists "Hilliard, Richard, Ironmonger and Timber Merchants" in New Street, Killarney. The Hilliard family was related by marriage to the Griffin family; Richard Hilliard married Rebecca Mansfield, the sister of Jack's mother.

By Christmas 1900 Jack had left his native country for good and embarked on a voyage that would take him to the other side of the world and to a new life in Australia.

The Mansfields

The Mansfields (see page 10) are believed to be of Norman origin, settling in Ireland in the 12th century[3]. Joseph Mansfield married Margaret Harman and they had ten children including John Harman Mansfield born in 1844 (Jack's Uncle) who emigrated to Australia around 1872 and settled in Maroonah Station, WA[3]. Joseph and Margaret Mansfield had a daughter Margaret born in 1849 (Jack's aunt), who married Daniel Dixon and lived in Enniscorthy, Co. Wexford, Ireland. In 1909, their son George Harman Dixon emigrated to Australia where he worked for a time on Maroonah Station[3].

Elizabeth (Jack's mother; see photograph page 5) was born in 1856 and Catherine (Jack's aunt) was born in 1859. Catherine married James Grey of Booterstown, Dublin, Ireland and they had a daughter Anne, who later went to Australia and lived on Maroonah Station with her uncle John Harman Mansfield[3].

New Street, Killarney in 2011. This is thought to be the site of R. Hilliard, Ironmongers where John Joseph Griffin worked for three years prior to emigrating to Australia in December 1900.

Rebecca (Jack's aunt) was born in 1861 and she married Richard Hilliard the ironmonger in Killarney in whose premises Jack worked for a number of years.

Irish Emigration

In 1890, when Jack was ten years old, two out of every five people born in Ireland were living outside Ireland[4]. Emigration had begun much earlier, as early as the 1600s, when people left Ireland for other parts of the British Isles, the British colonies, continental Europe and the islands of the Caribbean. According to some estimates ten million people have emigrated from Ireland since 1700 and today an estimated eighty million people worldwide claim some Irish descent[4].

Major destinations of Irish emigrants included Britain, the USA, Canada and Australia. Today, up to 25% of British citizens claim some Irish ancestry[5] while Irish Americans number over

45 million, making them one of the largest ethnic groups in the country[4]. In the 2006 census of Canada over four million Canadians claimed Irish descent and there are currently an estimated 500,000 people of Irish ancestry in Argentina[6]. In the 2001 census of Australia, Irish Australians numbered almost two million. The Irish also travelled to many other countries during the 19th and 20th centuries.

Due to a lack of deep water ports in Ireland the larger (and safer and cheaper) ships left from Liverpool or London; for this reason Irish emigrants would often go to Britain in order to travel further afield. In 1900 there were more than eighty cross-channel sailings per week to Britain from Dublin[7]. Kingstown (now Dun Laoghaire) was a point of departure for many emigrants to Britain and beyond.

References

[1] Ross, D. (2002), Ireland: History of a Nation, New Lanark: Geddes and Grosset, ISBN 1-84205-164-4.
[2] National Archives of Ireland.
http://www.census.nationalarchives.ie/exhibition/kerry/index.html
[3] Jones, A.E. (1998). Mansfield of Maroonah: From West Ireland to the Ashburton District of North West Australia 1874-1913 and beyond. Bibra Lake, WA. ISBN 0957788355.
[4] Wikipedia. http://en.wikipedia.org/wiki/Irish_diaspora
[5] The Guardian, Wednesday 13 September 2006.
http://www.guardian.co.uk/uk/2006/sep/13/britishidentity.travelnews
[6] Western People, Wednesday 14 March 2007.
http://www.westernpeople.ie/news/mhgbauojey/
[7] National Archives of Ireland.
http://www.census.nationalarchives.ie

Recommended Reading

• Jones, A.E. (1998). Mansfield of Maroonah: From West Ireland to the Ashburton District of North West Australia 1874-1913 and beyond. Bibra Lake, WA. ISBN 0957788355.

Chapter 2: Journey to Australia

Jack's Passage to Australia (1900)

In 1900 Queen Victoria made her fourth and final visit to Ireland, arriving in Kingstown (Dun Laoghaire) on the 4th of May. Jack was also in Kingstown that year with plans to leave his native Ireland for good. It is likely that he travelled by train from Kerry to Dublin and then by mail boat from Kingstown to Britain. We know from ships' passenger lists[1,2,3] that he left London on board the *SS Duke of Norfolk* on the 15th of December 1900, bound for Brisbane, Queensland, Australia. The journey took a full two months, despite taking the shorter route through the Suez canal.

The mail boat at Kingstown harbour c. 1900. Kingstown (now Dun Laoghaire) was a point of departure for many emigrants to Britain and beyond.

Outgoing passenger list for the SS Duke of Norfolk departing London 15th December 1900 bound for Brisbane, Australia. Note the entry for John Griffin age 21. From: The National Archives, London (http://www.nationalarchives.gov.uk and http://www.findmypast.co.uk)

Data from the incoming passenger list for SS Duke of Norfolk arriving Brisbane 16th February 1901. Original data from Queensland State Archives, Series ID 13086, Registers of Immigrant Ships' Arrivals, Rolls M471, M473, M1075, M1696-1710. From: http://www.ancestry.co.uk.

According to a newspaper article in the Brisbane Courier of the 18th February 1901, the *Duke of Norfolk* berthed in Brisbane arround eight in the morning of Saturday 16th February 1901.

The same newspaper of Monday 18th February 1901 describes the voyage from London to Australia[4]:

"The British-India steamer Duke of Norfolk, Captain Jenkyns, R.N.R., from London, via ports, arrived in Moreton Bay about midnight on Friday, and was assisted up the river yesterday to the Norman Wharf, where she berthed about 8 o'clock. She had a smooth passage from London, and the voyage proved most enjoyable. She left London on 15th December, called at Gravesend, where she picked up 475 immigrants and left the following day. Port Said was reached on the 29th of December, the weather being very fine, except for fogs off the coast of Portugal. She passed through the Canal on 30th December, left Suez the following day, and arrived at Colombo on 13th January. While the Duke was at Colombo 698 Boer prisoners arrived from Natal. She left Colombo the following day, calling at Batavia on 21st

[Top] VESSEL	DAY	MTH	YEAR	ARRIVED	DEPARTED
DUKE OF NORFOLK	05	02	1901	Townsville	....
HIOOTHA	09	02	1901	....	Hamburg
DUKE OF NORFOLK	13	02	1901	....	....
DUKE OF NORFOLK	13	02	1901	Mackay	....
DUKE OF NORFOLK	16	02	1901	Brisbane	London
DUKE OF WESTMINSTER S.S	27	02	1901	....	....
DUKE OF SUTHERLAND	03	03	1901	....	....
DUKE OF WESTMINSTER	04	03	1901	....	....
DUKE OF WESTMINSTER	07	03	1901	Brisbane	....
DUKE OF SUTHERLAND S.S	18	03	1901	....	....
DUKE OF SUTHERLAND	25	03	1901	Brisbane	London

Migrant ships arriving in Queensland 1837-1915. Original data extracted from microfilm copies of original records held at the Queensland State Archives in Runcorn, Brisbane. From: http://members.iinet.net.au/~perthdps/shipping/mig-qld5.htm)

The SS Duke of Norfolk seen here entering Cape Town during the Boer War.

January, and the usual Java ports. On the 31st of January the Duke arrived at Thursday Island, where she landed four immigrants and discharged 60 tons of cargo. The trip from Sourabaya to Thursday Island was negotiated in the good time of six days and five hours. At Cooktown she discharged 30 tons of cargo, and at Townsville 670 tons of cargo were landed and 175 Immigrants. It rained for three consecutive days at Townsville, which so delayed operations that the Duke was six days in port. A female immigrant died there of enteric fever, and was taken ashore for burial. The immigration agent at Townsville speaks very eulogistically of the immigrants landed there, who are said to be much above the usual class. At Mackay three immigrants were landed, and 50 tons of freight, and at Rockhampton 302 tons with fifty-six immigrants were landed, while at Maryborough sixteen passengers were dropped and 130 tons of cargo landed. For Brisbane the Duke of Norfolk has about 2000 tons of general cargo and 220 immigrants. The saloon passengers were Dr. and Mrs. Lyster for Townsville, and Trooper Starkey, of the Second Queensland Contingent. The latter made himself extremely popular on the voyage, and gave several magic lantern displays illustrative of scenes in the South African campaign, through which he had passed, while the immigrants should also have benefited by a lecture on life in Queensland. After discharging her cargo, the Duke of Norfolk is under orders for London, via the Cape."

Even for ships taking the "short" route through the Suez Canal voyages between Britain and Australia invariably took two months. The ships usually called into various ports *en route* and spent days in port loading and unloading cargo and supplies. The advent of container freight shipping in the 1950s and '60s changed all that. The time taken for loading and unloading was reduced to a matter of hours rather than days. Although this development revolutionised the shipping industry, led to greatly reduced transport costs and supported a vast increase in international trade, it was not popular with everybody. Dock workers, justifiably fearful for the future of their jobs, waged bitter industrial action and one ship's captain lamented that he used to be able to go ashore for a round of golf while his ship was unloading; now he barely has time for his lunch.

The Suez Canal

Jack travelled to Australia via the Suez Canal which links the Mediterranean Sea and the Red Sea. Opened in November 1869, approximately eleven years before Jack was born, it allows ships to travel between Europe and Asia without navigating around the Cape of Good Hope at the south of the African continent, a journey thousands of miles longer. The Mediterranean end is at Port Said and the southern terminus is at Suez. Interestingly, the longer route around the Cape has become popular again because of piracy in the Somalia region[5].

In recent years, the shrinking arctic ice has made the Northern Sea Route viable for commercial cargo ships on the route between Europe and East Asia for a short period in the summer months, a voyage thousands of miles shorter than the Suez Canal route. This route runs along the Russian Arctic coast and is often referred to as the Northeast Passage. According to climate researchers the

The Suez Canal c. 1900. Jack and the other passengers on board the Duke of Norfolk in late December 1900 would have gazed on a scene such as this.

route is likely to become passable without the help of icebreakers for a greater period each summer[6].

The Panama Canal that joins the Atlantic Ocean and the Pacific Ocean, another canal of great significance to international shipping, did not open until 1914, fourteen years after Jack emigrated to Australia.

The Boer War

While Jack Griffin was working in the ironmongers in Killarney, a bitter war had broken out in southern Africa between the Afrikaans-speaking Dutch settlers and the forces of the British Empire. The war was still in progress while Jack was travelling to Australia. Jack would have seen the Boer prisoners arriving in Colombo, Ceylon (Sri Lanka) as reported in the Brisbane Courier.

The Boer War was fought from 11th October 1899 until 31st May 1902. It ended with a British victory and eventual

The prisoner of war camp at Diyatalawa. Photo from angloboerwar.com

incorporation of the territory into the Union of South Africa, at that time a dominion of the British Empire. The conflict is commonly referred to as the Boer War but is also known as the South African War or the Anglo-Boer War. In the article on the arrival of the *SS Duke of Norfolk*, the Brisbane Courier refers to the conflict as "the South African campaign".

During the war, there was nowhere suitable in South Africa to hold the large numbers of Boer prisoners captured by the British forces. This led to the decision to send the prisoners away from South Africa. Of the 28,000 Boer men captured, 25,630 were sent overseas. The prisoners were sent to St. Helena, Ceylon (now Sri Lanka), Bermuda, Portugal and India. In Ceylon, a large camp was set up at the hill station at Diyatalawa, inland from Colombo. The hill station was reported to be good for the health of the incarcerated men and the prisoners were treated well[7]. Other camps of detention operated by the British during the Boer War did not enjoy such a good reputation[8].

Diyatalawa was almost certainly the destination for the 698 Boer prisoners from Natal that arrived in Colombo, Ceylon as witnessed by Jack and the other passengers and crew of the *Duke of Norfolk* in early 1901.

CEYLON BOER PRISON CAMP

Sherrill Babcock Says Captives Are Well Treated.

Their Officers, He Declares, Are Treat-ed as Equals by British Guards at Diyatalawa—Many of Them Paroled.

Sherrill Babcock, who has just returned from abroad, told yesterday of his visit and examination of the Boer prisoners' camp at Diyatalawa, Island of Ceylon, during the Summer of the past year. The three camps so far established by the British Government are at St. Helena, Ceylon, and in the Bermudas. About 5,000 prisoners are in the camp at Diyatalawa. Speaking of his visit, Mr. Babcock said:

" Diyatalawa is 162 miles from the coast and within about six degrees of the equator. Its altitude, however, is about 6,000 feet above the sea level, and the temperature is very agreeable. In the evenings, in fact, blankets are necessary.

" My first view of the Boer prisoners was obtained at Kandy. On my arrival at the Queen's Hotel I saw a group of four bearded gentlemen. On inquiring who they were I was told they were Boer officers, who were living at the Queen's Hotel, as they were permitted to leave the prison camp upon giving their parole.

A description of the prisoner of war camp at Diyatalawa from the New York Times of 2nd February 1902.

*Queensland Mounted Infantry, prior to departure to South Africa. Trooper Starkey
was a soldier in this division. Photograph courtesy of the John Oxley Library, State
Library of Queensland (Negative: 110279 http://nla.gov.au).*

Trooper E.R. Starkey, one of Jack's fellow passengers, was
one of the 10,000 Australian soldiers that fought with the British
forces in the Boer War[9]. Perhaps encouraged by his warm
reception on the *Duke of Norfolk*, trooper Starkey, of the Second
Queensland Contingent, went on to give further talks about his
war experiences. The Brisbane Courier of Wednesday 10th July
1901 reports a "Lecture on the War" to be given by ex-trooper
E.R. Starkey Q.M.I. of the Queensland South African Contingent
in the Alliance Hall, Woolloongabba (Brisbane) describing his
personal experiences and observations. *"The lecture-will be
illustrated by numerous limelight and bioscope views"*[10].

In 1901 people did not have access to cinema, television or the
radio. The first radio stations did not begin broadcasting until the
1920s (the first BBC radio broadcast from London was in 1922)
and television did not appear until many years later. Cinemas were
not common until the 1920s. Trooper Starkey's magic lantern,
limelight and bioscope shows would have been at the forefront
of modern entertainment and unusual enough to be worthy of a
mention in the newspaper.

The *SS Duke of Norfolk* (1889-1914)

In December 1900 Jack embarked on the *Duke of Norfolk* in London and travelled to Brisbane in Queensland[1,2,3]. The *Duke of Norfolk* (originally named the *Nairnshire*) was 3819 gross tons, 350 feet long and 48 feet wide[11,12]. She had a steam triple expansion engine, a single screw and a service speed of 10 knots. She was built in 1889 by R. & W. Hawthorn Leslie and Company, Limited (known simply as "Hawthorn Leslie") at the Hebburn Shipyard on Tyneside, England[11,12].

Vessel Name:	NAIRNSHIRE
Vessel ID:	515020021
Tonnage:	3,819
Owner:	Turnbull Martin & Co. Ltd.
Built:	1889
Date of Fate:	24 May 1914
Type of Fate:	Lost at sea
Region of Fate:	Rest of world
Vessel Abstract:	1889 NAIRNSHIRE in fleet of Turnbull Martin & Co. Ltd. 1889 sold to J.B. Westray, renamed DUKE OF NORFOLK. 1905 sold to Germany renamed MARCELLUS. 1908 sold to Sweden renamed JOHANNA. 1914 sold to Greece renamed PERICLES. 24 May 1914 foundered 90 miles from Ushant after striking submerged wreckage (Swansea for Alexandria).

SS Duke of Norfolk (formerly SS Nairnshire). Data from the New Zealand Marine News 1963 Volume 15 Number 2 Pages 8-10. From: http://www.nzmaritimeindex. org.nz

NAIRNSHIRE

Built by Hawthorn, Leslie & Company, Hebburn, England, 1889. 3819 gross tons; 350 (bp) feet long; 48 feet wide. Steam triple expansion engine, single screw. Service speed 10 knots.

Built for British owners, British flag, in 1889 and named **Nairnshire**. Turnbull, Martin, Glasgow. Sold to British owners, British flag, in 1899 and renamed **Duke of Norfolk**. Sold to Additional Arrivals, in 1905 and renamed **Marcellus**. Sold to Additional Arrivals, in 1908 and renamed **Johanna**. C.J. Banck, owner. Sold to Greek owners, in 1914 and renamed **Pericles**. C. Pappageorgacopulo, owner. Foundered in 1914.

SS Duke of Norfolk (formerly SS Nairnshire). From: http://www.ellisisland.org/ shipping/Formatship.asp?shipid=7458

The site of R. & W. Hawthorn Leslie and Co. Ltd. at Hebburn, Tyneside in 2007 (above) and in earlier days (below). http://www.tynetugs.co.uk/HawthornLeslie.html

Hawthorn Leslie was a shipbuilder and locomotive manufacturer founded in 1886. Perhaps the most famous ship built by the company was *HMS Kelly* launched in 1938 and commanded by Lord Louis Mountbatten. In 1968 the Company's shipbuilding interests were merged with two other yards to create Swan Hunter & Tyne Shipbuilders. The company's main shipbuilding yard at Hebburn closed in 1982, was sold to Cammell Laird and then acquired by A&P Group in 2001 but now lies derelict. Although Tyneside was once the third largest producer of ships in Britain, lack of modernisation, combined with competition from abroad, gradually caused the industry to decline and die.

The *Nairnshire* (*Duke of Norfolk*) was built for Turnbull, Martin & Co., a Glasgow cargo company founded in 1874. In 1884 the company began passenger services to New Zealand and in 1893 used the name Scottish Shire Line (not to be confused with the Shire Line of Steamers) for its ships sailing to New Zealand via South Africa and Australian ports.

The Eastern Steamship Company (established in 1871), or the Ducal Line as it was commonly known, operated a service between England and Calcutta, India[13]. In the 1890s the company, then controlled by the London shipbroker J.B. Westray & Co., began passenger services to Australia instead of India. In 1898 the company bought the *Nairnshire* from the Shire Line and renamed the ship the *Duke of Norfolk*[13,14].

Unfortunately for the Eastern Steamship Company, the Queensland trade was never as profitable as the original Calcutta business and the company was wound up in 1905[13]. The *Duke of Norfolk* was sold to a German company (C. Andersen, Hamburg)[14] and renamed *Marcellus*. Several years later, in 1908, a new Swedish owner (C.J. Banck)[12] renamed her the *Johanna*, and in 1914 she was sold to a Greek owner (C. Pappageorgacopulo)[12] and renamed *Pericles*[11,12,13,14]. On the 24th May 1914, just two months before the outbreak of World War I, she foundered 90 miles from Ushant at the Western end of the English Channel after striking submerged wreckage while on a voyage from Swansea in Wales to Alexandria in Egypt with a cargo of coal[11,13,14].

References

1 The National Archives, London
   http://www.nationalarchives.gov.uk
2 Queensland State Archives, Series ID 13086, Registers of Immigrant Ships' Arrivals, Rolls M471, M473, M1075, M1696-1710.
3 Migrant Ships Arriving in Queensland 1837-1915. Queensland State Archives, Runcorn, Brisbane.
4 The Brisbane Courier (Queensland), Monday 18th February, 1901.
5 The New American, 25th November 2008.
6 New York Times, 17th October 2011.
7 New York Times, 2nd February 1902.
8 The Concentration Camps.
   http://www.boer.co.za/boerwar/hellkamp.htm
9 National Archives of Australia.
   http://www.naa.gov.au/collection/explore/defence/conflicts.aspx#section1
10 Brisbane Courier, Wednesday 10th July 1901.
11 New Zealand Marine News (1963) 15(2): 8-10.
   http://www.nzmaritimeindex.org.nz
12 The Statue of Liberty-Ellis Island Foundation, Inc.
   http://www.ellisisland.org/shipping/Formatship.asp?shipid=7458
13 Ships in Focus, record 10, page 111. ISBN 1-901703-06-1
14 The Ships List.
   http://www.theshipslist.com/ships/descriptions/ShipsN.html

Chapter 3: Life in Australia

Arrival in Australia

Jack arrived in Brisbane, Queensland on Saturday the 16th of February 1901. He was on his way to Maroonah Station, a sheep station in the Ashburton District in the north-west of Western Australia, owned by his mother's elder brother John Harman Mansfield.

"Station" is the term for a large Australian landholding used for livestock production, similar to a ranch in North America. The owner of a station is called a grazier or pastoralist. Stations in Australia are, in most cases, on pastoral lease and are known as sheep stations or cattle stations as most have either one or the other dependent upon the land and rainfall. Sheep and cattle

Maroonah Station, 140 miles from Onslow. From the Western Mail (Perth) Saturday 25th December 1909.

Picking locks at Maroonah Station. From the Western Mail (Perth) Saturday 25th December 1909. After shearing, the wool is separated into categories: fleece, pieces, bellies, locks and crutchings.

stations can be thousands of square miles in area, with the nearest neighbour being hundreds of miles away.

An employee involved in routine duties on a station, including caring for livestock, is known as a station hand. A drover is a person, usually an experienced stockman, who moves livestock "on the hoof" over long distances. Reasons for droving include delivering animals to a new owner's property, taking animals to market, or moving animals during a drought in search of better food and water. Moving a small herd of animals is relatively easy, but moving several hundred head of wild station sheep or cattle over long distances is a completely different matter. A young person often works on a station for several years in a form of apprenticeship in order to become a Station Overseer (manager). Aborigines played a big part on many stations where they were considered to be competent stockmen.

Maroonah Station was a property of 212,300 acres situated 140 miles from Onslow and 220 miles from Carnarvon[1]. Brisbane, Jack's port of arrival in Australia in 1901, is a long way from

Western Australia. In those times, people travelled around the Australian continent on horseback, by horse-drawn transport or by coastal shipping services. Jack probably would have taken a coastal steamer or used an overland route. Air travel would not have been an option; the Wright brothers did not make their famous flight until December 1903, and the first airplane flight in Australia did not occur until 1910. Affordable mass-produced cars were not available until the 1920s. However by 1901, all the states except Western Australia were linked by railways, so it is possible that Jack travelled some of the journey by train.

Jack appears on the 1903 Australian Electoral Roll for the State of Western Australia, in the Swan district, Carnarvon subdistrict. He was by this time living on his uncle's station at Maroonah where he worked as a drover.

Western Australia is a very large and harsh land, especially the north-west where Maroonah is located. For a young man from the moist, green land of Co. Kerry, it must have been like arriving on a different planet. His work as a drover was in stark contrast to his job in an ironmongers in Killarney.

The Purchase of Maroonah Station

When John Harman Mansfield (Jack's uncle and owner of Maroonah station) died in 1907, Maroonah was left to his widow Annie Mariette Mansfield *née* McCafferty (1854-1911) with the proviso that Jack could buy the station from her, under the terms of the will. Jack borrowed the money and became the proud owner of Maroonah Station[2].

Jack's Marriage

The next official record of Jack in Australia is his marriage to Mary Glass on the 3rd of November 1909 in St. George's Church, Carnarvon "according to the rites and ceremonies of the Church of England". Mary Glass was the daughter of William Glass, a tailor, and Margaret Glass *née* Nicholl. On the marriage certificate Mary's age is given as 29 and Jack's age as 28. Jack

GASCOYNE WESTERN AUSTRALIA.

CERTIFICATE OF MARRIAGE.

Christian Names and Surname of the Parties	Age	Condition of the Parties (Spinster or Widow)	Rank or Profession or Occupation	Residence		Father's Christian Name and Surname	Rank or Profession of Father	Mother's Christian Name and Maiden Name
				Present	Usual			
John Joseph Griffin.	29	Bachelor	Engine Driver	Carnarvon	Mooramel Station Gullewon.	Edward Griffin.	Labourer	Elizabeth Maunsfield
Mary Glass.	24	Spinster	Candyf Compresson	Carnarvon	Gullewon Carnarvon.	William Glass.	Labourer	Margaret Nisbell.

I certify that the marriage of ....
....
a minor (or a) ....
a Justice of the Peace), was given to the Marriage between the
parties named in this Certificate.

This Marriage was celebrated between us } { John Joseph Griffin
Mary Glass.

In the presence of us } { John Glass
Rudy Miller

Marriage certificate of John Joseph Griffin and Mary Glass.

gives his profession as a "squatter". At that time in Australia the word squatter referred to a person of high social standing who grazed livestock on a large scale (whether the station was held by leasehold or freehold title). In Australia the term is still used to

Mary Griffin née Glass.

describe large landowners, especially in rural areas with a history of pastoral occupation[3]. Mary describes herself as a "lady's companion". A lady's companion was a woman of genteel birth who acted as a paid companion for women of rank or wealth[3]. One of the witnesses to the wedding was John Glass, a brother of Mary.

Mary Griffin *née* Glass

Mary was the second eldest child of William and Margaret Glass. She had seven siblings. The family originated from Ballycastle in Northern Ireland but the three oldest boys, as well as Mary, emigrated to Western Australia.

Some months after her marriage, Mary developed pneumonia and pleurisy. Very limited medical care would have been available in Australia at that time. In less populated parts of the country a pair of doctors could be expected to provide medical care for an area of almost two million square kilometres. The Royal Flying Doctor Service[4] which provides medical care for people in remote areas did not begin until many years later, in 1928. It is perhaps surprising then, that records indicate that Mary was attended by a doctor in the small town of Onslow in 1910 (it currently has a population of around 573 people[3]). However the choices for treatment would have been limited. Antibiotics, for example, were not available in 1910; they were not developed until the 1940s.

Sadly, Mary died of pneumonia and pleurisy aged 31 years on the 11th of April 1910 in the Rob Roy Hotel, Onslow. This was the Old Onslow Township which is south of the modern Onslow. The doctor had attended her for two weeks before she died. Mary was buried on the same day she died (which was normal at that time) at the Onslow Cemetery, now referred to as the Old Onslow Cemetery. At the time of her death Mary had been in Australia for three years and had been married for less than six months. Recollections of family members[2] suggest that Mary may have been pregnant at the time of her death although there is no official verification of this.

The Old Onslow Cemetery in Western Australia.

Tombstone of Mary Griffin née Glass at the Old Onslow Cemetery in Western Australia.

THE ONSLOW HURRICANE
BOATS DRIVEN ASHORE CARRIED HALF A MILE INLAND
Perth, April I8.
Further particulars concerning the recent disastrous blow at Onslow, which lasted 19 hours, state that all the boats broke away from their moorings, and the *Collier*, a 75-ton boat, laden with cargo, was carried a quarter of a mile inland. The *Doris* was turned upside down on the beach. Four luggers, with crews of 24, comprising Japs [sic] and Malays, were lost during the storm, and a fifth was carried half a mile inland. The parish hall was completely demolished, and the roofs of the Rob Roy Hotel and Clarke's store were carried away. Other buildings were also more or less damaged.

The Onslow Hurricane. From: The Advertiser, Monday 19th April 1909

The original town of Onslow (Old Onslow, also known as Ashburton or Old Ashburton), was abandoned fifteen years later, around 1925, due to its vulnerability to hurricanes[5] and also to repeated difficulties with the mouth of the Ashburton River silting up. A new town was built on the other side of the bay about 46 km away from the old town[6].

DROUGHT IN WESTERN AUSTRALIA.

PERTH.— In consequence of the shortage of water in drought-stricken country districts, the Government has arranged to truck supplies to railway stations and sidings within the areas.

Drought in Western Australia. From: The Sydney Morning Herald, Monday 5th October 1914, page 4.

The Loss of Maroonah

Jack had acquired Maroonah in 1907 by borrowing approximately £5000 which would be the equivalent of around £400,000 today (2011)[7].

Between 1911 and 1916, Australia suffered a major drought with the resultant loss of nineteen million sheep and two million cattle[8,9]. By 1913 Jack's debts stood at just under £9000 and he clearly realised that he could not continue running the station. He put the property up for auction in early 1913[1]. The advertisement in the West Australian of Wednesday 8th January 1913 reads as follows:

MAROONAH STATION, CARNARVON.
212,300 ACRES, 10,800 SHEEP, 70 HORSES and PLANT.
UNITED SERVICE HOTEL, PERTH.
On MONDAY. FEBRUARY 10. 1913. At Three o'Clock.
ELDER, SHENTON and CO., LIMITED (in conjunction with Angelo and Angelo) have received instructions from Mr. J. J. Griffin to OFFER by AUCTION as above, his well-known MAROONAH STATION Containing 212,300 acres of Pastoral Lease, situated 140 miles from Onslow and 220 miles from Carnarvon. The boundary is enclosed by 127 miles of 6-wire fencing in good order, the property being subdivided into 9 sheep proof paddocks, and there are the usual station buildings, stockyards, shearing shed, etc. Water. The property is watered by 6 pools and springs, in addition to 9 wells, on six of which are erected windmills, tanks, and troughing, stock include about 10,800 sheep, consisting of about 6,600 ewes, 3,0000 wethers, two and four years old, in good condition, 1,000 lambs and 200 rams.

Seventy head of light and draught horses, 2 waggons, harness, blacksmith's tools, and all necessary plant for carrying on the station.

AUCTIONS.

MAROONAH STATION, CARNARVON.

212,300   ACRES.   212,300.

10,800 SHEEP, 70 HORSES and PLANT.

UNITED SERVICE HOTEL, PERTH.
On MONDAY, FEBRUARY 10, 1913.

At Three o'Clock.

ELDER, SHENTON and CO., LIMITED (in conjunction with Angelo and Angelo) have received instructions from Mr. J. J. Griffin to OFFER by AUCTION as above, his well-known

MAROONAH STATION.

Containing 212,300 acres of Pastoral Lease, situated 140 miles from Onslow and 230 miles from Carnarvon.

The boundary is enclosed by 127 miles of 6-wire fencing in good order, the property being subdivided into 9 sheep proof paddocks, and there are the usual station buildings, stockyards, shearing shed, etc.

Water.—The property is watered by 6 pools and springs, in addition to 9 wells, on six of which are erected windmills, tanks, and troughing, stock include about 10,800 sheep, consisting of about 6,600 ewes, 3,000 wethers, two and four years old, in good condition, 1,000 lambs and 200 rams.

Seventy head of light and draught horses, 2 waggons, harness, blacksmith's tools, and all necessary plant for carrying on the station.

All information from
ELDER, SHENTON and CO., LIMITED
Perth.

Or

ANGELO and ANGELO,
Carnarvon.

Auction of Maroonah Station. From: The West Australian (Perth), Wednesday 8th January 1913, page 3. The advertisement for the auction of Maroonah Station to be held on Monday February 10th, 1913 in Perth, Western Australia.

The auction was not a success and Jack tried again to sell the station at another auction in May. It appears, however, that nobody wanted to buy a sheep station in Western Australia during one of the worst droughts for decades. The bank had stopped Jack's credit in March and in August they moved to seize his property. The Kalgoorlie Western Argus of Tuesday 30th September 1913 reports his bankruptcy[10]:

> *"Bankruptcy Notices: Among the gazetted notices under the Bankruptcy Act are: Adjudications: Edward Ashenden, Boulder, greengrocer; William Ager, Mayndale, Bridgetown, farmer. Receiving Orders: Edward Heads, Three Springs, farmhand; William Ager, Mayndale, Bridgetown, farmer;* **John Joseph Griffin**, *Olivia Terrace, Carnarvon, lately residing and carrying on business at Maroonah Station in the Gascoyne and Ashburton districts, pastoralist."*

The first meeting and public examination of J. J. Griffin was set for October 14th (announced in The West Australian of Saturday 27th September 1913). An account of the bankruptcy hearing held on October 28th was reported by The West Australian of Wednesday 29th October 1913:

IN BANKRUPTCY. TUESDAY, OCTOBER 28.
Re J. J. Griffin.

John Joseph Griffin said he was a pastoralist, of Carnarvon. He owed to 15 unsecured creditors £4,063, and £4,800 to a fully-secured creditor. His securities he estimated to be worth £6,750, in addition to which he had a half-share in some Fremantle property, valued at £300; a consignment of wool which he had sent to London, and which he valued at £526; and a small piece of land in Onslow, valued at about £20. It was in March last, when he owed nearly £5,000, that the bank stopped his credit. But for a big loss in sheep, owing to dry seasons, he would have occupied a

different position. The bank seized in August.
The Official Receiver: *When does the lease of the station expire?*
Debtor: *I have never gone into that.*
The Official Receiver: *It surprises me to think you do not know what the tenure is of a proposition involving such big figures.*
Debtor: *I did not think there was any fixed time for the lease to run out.*
Debtor, continuing, *said the station had been a losing proposition for the last two or three years.*
The Official Receiver: *You were fortunate enough to receive a very excellent benefit under your uncle's will and to acquire your interest in this station on exceptionally simple terms, yet within a very short time your splendid heritage is gone and there is little prospect of your creditors being paid. Is this position due only to droughts?*
Debtor: *Yes.*
The examination was adjourned.

In the West Australian of Wednesday the 17th September 1913 we find a notice for the auction of Maroonah Station "under instructions from the mortgagee" to be held on Thursday 2nd October in Perth, Western Australia. The bank had foreclosed and Jack was no longer living on the station, at this time giving his address as Olivia Terrace, Carnarvon rather than Maroonah. It is hard to imagine the effect the loss of his beloved Maroonah would have had on Jack, especially coming so soon after the death of his wife.

Jack was an excellent horseman and rifle shot and expert with a stockwhip[2]. He got a job as a drover on neighbouring Mulga Downs Station, which was owned by George Hancock (1882-1960), the father of future iron ore mining magnate, Lang Hancock (1909-1992)[2].

At this time, Jack's sister Madge and her son Tom were living on the coast at Onslow, where she met and subsequently married in 1913 a young English Methodist minister Edward Jones (1890-1963). By late 1915 she was living in White Road, Bunbury, Western Australia and by May 1916 she had moved to Albany, Western Australia[11].

During World War I Jack volunteered for service with the Australian Imperial Force (AIF)[11]. He served in Australia, Egypt and France and was wounded during active service in the Battle of the Somme.

References

[1] The West Australian (Perth), Wednesday 8th January 1913, page 3.
[2] Jones, A.E. (1998). Mansfield of Maroonah: From West Ireland to the Ashburton District of North West Australia 1874-1913 and beyond. Bibra Lake, WA. ISBN 0957788355.
[3] Wikipedia. http://en.wikipedia.org
[4] Royal Flying Doctor Service. http://www.flyingdoctor.org.au
[5] The Advertiser, Monday 19th April 1909.
[6] Old Onslow.
http://www.chapelhill.homeip.net/FamilyHistory/Photos/Old_Onslow-WA/
[7] RPI method. http://www.measuringworth.com/ukcompare/
[8] Australian Government Bureau of Meteorology.
http://www.bom.gov.au/climate/drought/livedrought.shtml
[9] The Sydney Morning Herald, Monday 5th October 1914, page 4.
[10]The Kalgoorlie Western Argus of Tuesday 30th September 1913.
[11]Next of Kin addresses provided in AIF Service Record (see Appendix I). National Archives of Australia. Search for "Griffin 4797" at http://naa12.naa.gov.au/

Recommended Reading

• Jones, A.E. (1998). Mansfield of Maroonah: From West Ireland to the Ashburton District of North West Australia 1874-1913 and beyond. Bibra Lake, WA. ISBN 0957788355.

- Valli, Jack (1983). Gascoyne Days. St. George Books, Perth, WA. ISBN 0867780193.
- Forrest, K. (1996). The Challenge and the Chance: The Colonisation and Settlement of North West Australia 1861-1914. Hesperian Press, Carlisle, WA. ISBN 0859052176.

Chapter 4: World War I

World War I

About nine months after Jack left Maroonah Station, thousands of miles away in a country known today as Bosnia-Herzegovina, Archduke Franz Ferdinand, the heir to the throne of Austria-Hungary, was assassinated by a Serbian nationalist in the city of Sarajevo. This event triggered the conflict now referred to as World War I or the Great War[1]. This war lasted from the 28th July 1914 until the 11th November 1918 and ended in an Allied victory. By the end of the war four empires (German, Russian, Ottoman, and Austro-Hungarian) had ceased to exist and the map of central Europe had been completely redrawn. More than nine million combatants were killed during the conflict[1].

For Australia, as for many nations, World War I was the most costly conflict in terms of deaths and casualties[2]. At that time Australia had a population of less than five million people. From this small population 416,809 men enlisted voluntarily. During the war 60,000 of these men were killed and 156,000 wounded (of which Jack was one), gassed, or taken prisoner[2].

1st Australian Imperial Force (AIF)

The Australian Imperial Force (AIF) was the name given to the military forces raised by Australia in World War I[3]. Under the provision of the Defence Act 1903, enlistment for service overseas was voluntary. During World War I, Australia and South Africa were the only countries in the war which did not resort to conscription. The AIF was disbanded on the 1st April 1921[3], although a Second Australian Imperial Force (2nd AIF) was formed during World War II for volunteers in the Australian Army.

Enlistment in the AIF

The outbreak of war was greeted in Australia, as in many other places, with great enthusiasm[2]. The number of people volunteering to enlist in the AIF was so high that recruitment officers had to turn people away. However, as the war went on and casualty rates increased, the number of volunteers declined. Jack was one of the 3,000 Irish-born Australians that volunteered for service in the AIF [4].

John Joseph GRIFFIN

Regimental number	4797
Place of birth	Kilarney, Ireland
Religion	Church of England
Occupation	Drover
Address	Roebourne, Western Australia
Marital status	Single
Age at embarkation	35
Height	5' 8"
Weight	150 lbs
Next of kin	Sister, Mrs M C Jones, c/o Ezy Watkins Ltd, York Street, Albany, Western Australia
Previous military service	Nil
Enlistment date	29 December 1915
Place of enlistment	Blackboy Hill, Western Australia
Rank on enlistment	Private
Unit name	11th Battalion, 15th Reinforcement
AWM Embarkation Roll number	23/28/4
Embarkation details	Unit embarked from Fremantle, Western Australia, on board HMAT A38 *Ulysses* on 1 April 1916
Rank from Nominal Roll	Private
Unit from Nominal Roll	51st Battalion
Fate	Returned to Australia 13 February 1917
Discharge date	11 June 1917

Embarkation roll data for John Joseph Griffin of the Australian Imperial Force (AIF)[5].

A 14720. **TRANSFERRED TO** 51st Battalion

AUSTRALIAN MILITARY FORCES.

AUSTRALIAN IMPERIAL FORCE

Attestation Paper of Persons Enlisted for Service Abroad.

No 4997 Name *GRIFFIN JOHN JOSEPH*

Unit *9 Depot*

Joined on *29-12-15*

Questions to be put to the Person Enlisting before Attestation.

1. What is your Name?	1. *John Joseph Griffin*
2. In or near what Parish or Town were you born?	2. In the Parish of ... in or near the Town of *Killarney* in the County of *Ireland*
3. Are you a natural born British Subject or a Naturalized British Subject? (N.B.—If the latter, papers to be shown.)	3. *N.B.*
4. What is your age?	4. *34 yrs 10 mths*
5. What is your trade or calling?	5. *Drover*
6. Are you, or have you been, an Apprentice? If so, where, to whom, and for what period?	6. *3 yrs R Hillard Coonamble New to Killarney Ireland*
7. Are you married?	7. *No*
8. Who is your next of kin? (Address to be stated)	8. *Sister Mrs E Jones White St Bunbury W.A.*
9. Have you ever been convicted by the Civil Power?	9. *No*
10. Have you ever been discharged from any part of His Majesty's Forces, with Ignominy...?	10. *No*
11. Do you now belong to, or have you ever served in, His Majesty's Army...?	11. *No*
12. Have you stated the whole, if any, of your previous service?	12. *Yes*
13. Have you ever been rejected as unfit for His Majesty's Service?	13. *No*
14. (For married men...) Do you understand that no separation allowance will be issued...?	14.
15. Are you prepared to undergo inoculation against small pox and enteric fever?	15. *Yes*

I, *John Joseph Griffin* do solemnly declare that the above answers made by me to the above questions are true, and I am willing and hereby voluntarily agree to serve in the Military Forces of the Commonwealth of Australia within or beyond the limits of the Commonwealth.

And I further agree to allot not less than two-fifths of the pay payable to me from time to time during my service for the support of my wife and children.

Date *29-12-15*

Signature of person enlisted.

Enlistment of John Joseph Griffin in the Australian Imperial Force (AIF) December 1915[6].

Blackboy Hill today. The memorial and flagpole are aligned for the setting sun on Anzac Day (photo by SatuSuro).

At the age of 34, Jack enlisted as a private at Blackboy Hill, Western Australia[5,6]. Blackboy Hill was a military training camp used to house large numbers of AIF troops before they left for war. The adjacent Helena Vale Railway Station (now called Midland Junction) was used to transport the troops to Fremantle, where they boarded troopships to take them to the battlefields of Europe and the Middle East.

A medical examination of John Joseph Griffin undertaken at Roebourne, Western Australia on the 4th December 1915 describes Jack as 5 ft. 8 ins. tall, 150 lbs. (or possibly 15 st., the wording is ambiguous) with a chest measurement of 46 ins. The medical report notes that he had dark brown hair, light blue eyes and a "fresh" complexion (see page 47)[6].

3

Description of *John Joseph Griffin* **on Enlistment.**

Age ...34... years ...10... months.

Height ...5... feet ...8... inches.

Weight ...15.0... lbs.

Chest Measurement ...46... inches.

Complexion ...fresh...

Eyes ...L. Blue...

Hair ...D. brown...

Religious Denomination ...C.E...

DISTINCTIVE MARKS.

nil

CERTIFICATE OF MEDICAL EXAMINATION.

I HAVE examined the above-named person, and find that he does not present any of the following conditions, viz.:—

Scrofula; phthisis; syphilis; impaired constitution; defective intelligence; defects of vision, voice, or hearing; hernia; hæmorrhoids; varicose veins, beyond a limited extent; marked varicocele with unusually pendent testicle; inveterate cutaneous disease; chronic ulcers; traces of corporal punishment, or evidence of having been marked with the letters D. or B.C.; contracted or deformed chest; abnormal curvature of spine; or any other disease or physical defect calculated to unfit him for the duties of a soldier.

He can see the required distance with either eye; his heart and lungs are healthy; he has the free use of his joints and limbs; and he declares he is not subject to fits of any description.

I consider him fit for active service.

Date ...4 Dec 15...

Place ...Roebourne...

I. Maunsell

Signature of Examining Medical Officer.

Medical examination of John Joseph Griffin undertaken at Roebourne, Western Australia 4th December 1915 shows Jack was 5 ft. 8 ins. tall, 150 lbs. (possibly 15 st.) with a chest measurement of 46 ins. He had dark brown hair, light blue eyes and a "fresh" complexion.

Army Form B. 103.

Casualty Form — Active Service.

Regiment or Corps ____   3RD TRAINING BATTALION   "C" COMPANY

Rank   Pte   797   Name ____

Terms of Service (a) ____

to ____ Re-engaged

Date of appointment } to lance rank

Service reckons from (a) ____

Numerical position on } roll of N.C.O's.

Qualification (b)

Record of promotions, reductions, transfers, casualties, &c., during active service, as reported on Army Form B 213, Army Form A. 36, or in other official documents. The authority to be quoted in each case.	Place	Date	Remarks taken from Army Form. B. 213 Army Form A. 36) or other official documents.
... at	SUEZ Alexandria	25.4.16	A.P 7372 ex Ulysses D7517
Embarked	Alexandria	7.6.16	A P 8298
Disembarked	Marseilles	14.6.16	L R 5811
	Etaples	6.7.16	A P 1872
		6.7.16	A P 179
		6.7.16	A.P

Casualty Form-Active Service 1. Service record indicating Jack's arrival in Alexandria on the 25th April 1916, the embarkation for France on the 7th June and his subsequent arrival in Marseilles. Also noted is the treatment he received for a sprained ankle in Étaples in July, 1916.

CERTIFICATE OF COMMANDING OFFICER.

I CERTIFY that this Attestation of the above-named person is correct, and that the required forms have been complied with. I accordingly approve, and appoint him

to,...

28 FEB 1916

Date...

Lt. Col.,

Camp Commandant

Appointment to battalion. Following his military training at Blackboy Hill, Jack's commanding officer appoints him to his unit and Jack is sent to Fremantle to embark on HMAT Ulysses.

Jack's service record shows that he enlisted on 29th December 1915 in the 11th Battalion (15th Reinforcement)[5,6]. It is thought that he had tried to enlist at the outbreak of war but was refused on the grounds that he was too old. However, by December 1915 the army needed more volunteers and were less selective. Following his military training at Blackboy Hill, Jack, along with 203 other soldiers of the 11th Battalion, 15th Reinforcement, embarked on His Majesty's Australian Transport *HMAT A38 Ulysses* on the 1st April 1916. One of these soldiers, Herbert Charles Cruttenden of Eden Hill, West Guildford, Western Australia, kept a personal diary[7]. In his diary, Private Cruttenden describes the journey from Fremantle to Alexandria via the Suez canal and then to a training camp at Tel el Kebir in Egypt. From Tel el Kebir, Jack and his regiment were sent to the front line in France:

"We left Fremantle on April 1st 1916 and landed in Suez on April 23rd 1916, it was the best trip they had ever had for years. We saw the first A on entrance to the Canal; we went through the Canal to Alexandria and landed there on April 25th 1916 and proceeded to Tel el Kebir camp where we stayed for seven weeks and then went to France. We landed in Marseilles on June 12th 1916 and went to our camp at a place called Étaples, about twenty miles from Boulogne. A training camp for all soldiers because it is

HMAT Ulysses. Photograph from 1st March 1916[8].

> *the nearest and best to get to the firing line, about seventy*
> *miles from the firing line."*

Jack's "Casualty Form-Active Service" (see page 48)[6] confirms
the details of the journey as recorded by Private Cruttenden[7].

HMAT A38 Ulysses

A fleet of transport ships was leased by the Australian government
for the specific purpose of transporting the AIF troops to
war[8]. When not being used for military transport, these ships
were employed to carry cargo to Britain and France. The fleet
was composed of British ships and captured German vessels.
Launched in 1913, the *Ulysses* was the largest ship to serve as
a troop carrier. She weighed 14,499 tons with an average cruise
speed of 14 knots or 25.92 kph[8]. The *Ulysses* was owned by the
China Mutual SN Co., London and leased by the Commonwealth
until the 15th August 1917. She also sailed between Australia and

Europe during the Second World War, again ferrying Australian troops and airmen to the front. The *Ulysses* was torpedoed by an unknown German submarine in 1942 and sunk off Florida after apparently disobeying an order that would have led her through safer waters[8].

Arrival in Alexandria

During the long voyage from Australia the men would have been plagued not only by the misery of seasickness but also by anxiety about German U-boats which were at that time targeting Allied troop carriers. By the time they arrived in Egypt, the troops had been in *HMAT Ulysses* for a considerable period and must have been glad to reach their destination. Major T.W. Edgeworth David of the Australian Mining Corps[9] (known by the nickname of "the Tunnellers") who had shared the voyage aboard the *Ulysses* with Jack's regiment, describes the behaviour of some of his men when the *Ulysses* eventually arrived in the Egyptian port of Alexandria on the 25th April 1916[9]:

> *"On the arrival of our troop ship in Alexandria in April 1916, a party of some 120 out of our 1200 miners, with the wanderlust strong upon them, broke loose suddenly from our troop ship as she lay at the wharf, rushed the sentries, and went careering like a lot of released school boys up the main street of Alexandria, making for the heart of the city. Some bad sport, perhaps one should rather say, one sound disciplinarian, telephoned to the military police."*

It appears that the men were rounded up and spent the night in a prison on shore. However, in the early dawn, Major David received an agonized SOS from the military police to say that the soldiers were tunnelling under the walls, and that the prison

> *"was tottering to its foundations, and would we send up a strong-armed party at once to hold and remove the prisoners".*

Habeita. Carinya Camp, a Suez Canal defence post on the front line about twelve miles east of the canal. There is a shelter for horses on the left. Rows of rocks separate the lines of eight person Indian production (EPIP) tents[10].

Tel el Kebir Camp

On arrival in Alexandria, Jack's regiment was sent to Tel-el-Kebir military camp, located 75 kilometres south of Port Said on the edge of the Egyptian desert[11]. Tel el Kebir was a training centre for the AIF reinforcements. Around 40,000 Australians were camped at Tel-el-Kebir in a "tent city" about six miles in length. There was a military railway to transport troops between the camp and their vessels in Alexandria. One soldier, Ernest George King, describes Tel el Kebir in his personal diary[11] as *"a very dirty little place with a few dirty shops in it"*.

The 51st Battalion

Jack had enlisted in the 11th Battalion. However, in Egypt during the first week of March 1916 a new battalion, the 51st, was formed as part of a reorganization of the AIF. Approximately half of its recruits were Gallipoli veterans from the 11th Battalion and the other half were fresh reinforcements from Australia; Jack was one of the latter. Like the 11th Battalion, the 51st was predominantly composed of men from Western Australia. The battalion became part of the 13th Brigade of the newly-formed 4th Australian

Trenches on the Western Front.

Division[12]. Jack and his battalion spent some time manning the Suez canal defences at Habeita before they were shipped to the battlefields of France[7].

On the 1st of June the battalion received orders for France. The next three days were spent in refitting and equipping. On the 4th of June the Battalion left Serapeum West Siding in two trains of open trucks bound for Gabbury Quay at Alexandria, where Jack and his regiment embarked for France. Within a fortnight of arriving in France Jack was in the trenches of the Western Front[7].

Battle of the Somme

The Battle of the Somme, or the Somme Offensive as it is also known, refers to a series of battles fought between the 1st July and the 13th November 1916 along the Somme Valley in France. On the first day of the offensive, 1st July 1916, the British army suffered almost 60,000 casualties, a third of whom were killed[13]. The word "Somme" has since become synonymous with battlefield slaughter. By the time the offensive was abandoned in November, the allied forces had managed to advance only twelve kilometres. The battle resulted in around 500,000 German casualties from which the German army never recovered; however, this had come at a cost of 620,000 Allied casualties[13]. The major contribution of the Australian troops to the Somme offensive was in the fighting around Pozières and Mouquet Farm between 23rd July and 3rd September[13].

Jack spent little time in Marseilles before beginning the sixty hour train journey north. The railway tracks run straight through the Rhone Valley and the train stopped every four hours so the soldiers could be given refreshments donated by the local citizens. The battalion arrived at Caestre not far from the Belgian border and went into billets at Moolenacker. Moolenacker was close to the front line, close enough to hear the distant rumble of the heavy guns and the sound of the occasional aircraft fire, horrifying in its intensity[7].

Ten days were spent training for trench warfare and on the 26th of June they were in the Petillon Section near Estaires. Jack's battalion suffered its first casualties and by the 30th of June five men had been killed, another died of wounds and a further seventeen had been wounded[7].

The battalion spent most of July and the early part of August in exercises in the morning with rest in the afternoon. Gradually the battalion moved towards the frontline via Herissart, Valdencourt and Albert until, on the 13th of August, they found themselves in Wire Trench near Ferme de Mocquet, or Mouquet Farm[7]. Jack's battallion fought in its first major battle at Mouquet Farm on the 14th August 1916 and suffered casualties equivalent to a third of its strength[7].

The Battle of Mouquet Farm

"Mouquet Farm" is the name given to a series of Australian attacks northwards along the Pozières Heights between the 8th of August and the 3rd of September 1916. They followed on from the seizure of Pozières and the German lines at the windmill east of the village in late July and early August[14,15].

The battle for Mouquet Farm, just north of Pozières, is still shrouded in controversy. Thousands of Australian troops died

Mouquet Farm before its destruction by shellfire in 1916[16,17].

over a period of several weeks while the farm was taken and abandoned a number of times[15]. In the fighting at Pozières and Mouquet Farm the Australians suffered a loss comparable with the casualties sustained over eight months at Gallipoli in 1915[14].

Mouquet Farm after its destruction by shellfire in 1916[18]. The farm buildings were reduced to rubble although strong stone cellars remained below ground which were incorporated into the German defences.

Upper Bath

Army Form B. 179.

4366

Medical Report on an Invalid.

Station _____ *[stamp: No. 2 AUSTN. COMMAND DEPOT A.M.C. 10 NOV 1916 WEYMOUTH]*

Date _____

1. Unit 51st Battn. A.I.F.
2. Regimental No. 4191
3. Rank Private
4. Name GRIFFIN John Joseph

5. Age last birthday 35
6. Enlisted { on 29th December 1915 at York
7. Former Trade or Occupation { Station Overseer

8. Disability.

Gun shot wound right upper arm and forearm.

Statement of Case.

Note.—The answers to the following questions are to be filled in by the Officer in medical charge of the case. In answering them he will carefully discriminate between the man's unsupported statements and evidence recorded in his military and medical documents. He will also carefully distinguish cases entirely due to venereal disease.

9. Date of origin of disability. 14th August 1916

10. Place of origin of disability. Pozières France

11. Give concisely the essential facts of the history of the disability, noting entries on the Medical History Sheet bearing on the case.

S.W. right forearm + arm Aug 14th. Injury to median nerve.

12. (a) Give your opinion as to the causation of the disability. Shell wound

(b) If you consider it to have been caused by active service, climate, or ordinary military service, explain the specific conditions to which you attribute it (See notes on page 3). Active service

Jack's service record (see page 56) states that he was wounded in action by a gunshot wound (or shell wound) to the right upper arm and forearm on the 14th August 1916 at Pozières, France while on active service[6]. Injury to the median nerve is also noted. The medical report is dated the 10th of November 1916 at Weymouth, England[6].

Sergeant Percy Nuttall of the 50th Battalion who came from Victoria, Australia, kept a personal diary of events of the Battle of Mouquet Farm[15]. His entry for the 14th August reads as follows:

"14th August 1916: Rain and mud galore. No sleep, little drink and nothing to eat. Wounded craving for a drink, but not many casualties until 3pm when Fritz turned his artillery on and he did stir us up and wrecked our trenches. My platoon, who I was in charge of, lost heavily and about 5 o'clock 12 Platoon was handed over to me sixteen strong. They having only one NCO, one L/Cpl left. W Riches and Musgrave were sent to hold a shell hole and are dead. [Someone] told me a shell burst either on or over them. After some persuasion the Major let me go over to see what happened and there I found Riches dead and Musgrave a pitiful sight under Riches, shell shocked and smothered with his mate's blood.

About 7pm, orders came that we had to make another advance at 9.30 and dig in between a quarry and Mouquet Farm. Major Herbert wrote back to headquarters that we were not strong enough to undertake the job. They replied it had to be done at all costs. So at the given time we moved out with our 300 men. Headquarters got to know Fritz was going to attack us, and soon as we moved we got full force of their barrage which killed or wounded half our strength. I got one in the ribs, and one half of my body went numb, but I heard the Captain say 'follow on C Company', so I went and took up our position after trying to rally the lads together. When we had dug in about three feet, word came that we had to retire as the battalions on our left and right did not join up, which was heart breaking.

I was told to go over to the right flank to take charge. There I went only to find confusion as the lads did not know how far to go, so I called for the bombers and only one responded. So the two of us went down the trench, me with the bayonet and Tom Ryan with the bombs but only ran across a platoon of A Company who were challenged and luckily let in as they came in from 'no man's land'. It proved afterwards they had got lost. We stood to the rest of the night and only Fritz's patrols were seen, but we kept them off at daylight. I was told off to count the battalion, which comprised 156 men and three officers unwounded. I was then put on rationing them and the sights I saw is indescribable. We tried all day to get the wounded back but Fritz's fire delayed operations."

The next section mentions an attack on Jack's battalion, the 51st:

"Burying the dead was impossible as they still kept up a terrible fire from high explosives – howitzers and whizz bangs. During the afternoon word came of our relief as were too weak to hold out any longer. After all arrangements the 4th Battalion came along and we started to work our way out. All went well until 9.30pm when going through the 51st Battalion trench the rotters sighted us and sent up three green lights which I knew the meaning immediately. I found myself buried under 5 feet of earth, with my neck bent on my chest, near my knees in a sitting position, my hand clasping my rifle. Another chap Stevens in the 51st Battalion had my leg bent over him. His head was near my feet. I sang out a few times and then settled to my fate, which was a glimpse of the past. I also swore vengeance on the Huns, saying they had three solid days to get me fair and square and couldn't - they had to pick this rotten

way. On top the relief party was doing its best, the Huns'
machine guns playing on them all the time. After an hour
I felt my equipment gripped, when I was on my last breath
due to my awkward position, up came my head and what
a relief. I immediately freed my arms and set to work
burrowing for the chap near my feet, knowing where his
head was. Up he came ten minutes later. We shook hands,
of course, pleased we were."

The "Stevens" of the 51st battalion mentioned in the last paragraph of Sergeant Nuttall's account is probably Alexander James Stevens who was a stockman from Queensland, Australia[19]. He would almost certainly have known Jack and fought alongside him. Pte. Stevens returned to Australia on the *HMAT Benalla* on the 25th August 1917 suffering from shell shock.

Sergeant Percy Nuttall, to whom we are indebted for his graphic account of life in the trenches during the Battle of Mouquet Farm, was later promoted to lieutenant and awarded the military cross. He was wounded in action in the second battle of Villers-Brettonneux and returned to Australia where he lived until his death in 1966[15].

We are also indebted to Herbert Charles Cruttenden of Eden Hill, West Guildford, Western Australia for his account of the 51st Battalion and the fighting at Pozières and Mouquet Farm[7]. Sadly, in early September, still fighting over Mouquet Farm, 160 men of the 51st Battalion were surrounded and never seen again. Eight years later officers of the War Graves Commission found a trench full of Australian bodies just beyond Mouquet Farm. Private Cruttenden was one of the Australian soldiers who never returned from the war.

References

[1] Wikipedia. http://en.wikipedia.org/wiki/World_War_I
[2] Australian War Memorial. First World War 1914–18.
  http://www.awm.gov.au/atwar/ww1.asp

[3] Australian War Memorial. 1st Australian Imperial Force.
http://www.awm.gov.au/units/unit_13200.asp

[4] Irish-born enlistments in the AIF.
http://www.aif.adfa.edu.au/examples.html#Irish

[5] The AIF Project.
http://www.aif.adfa.edu.au:8080/showPerson?pid=119767

[6] National Archives of Australia. Search for "Griffin 4797"at
http://naa12.naa.gov.au/

[7] Personal Diary of Herbert Charles Cruttenden.
http://homepage.ntlworld.com/ian.cruttenden1/military_
service/10779.htm

[8] *HMAT Ulysses.* http://25thlondon.com/ulysses.htm

[9] In Great Haste. Edgeworth David: The 'Knight in the Old
Brown' Hat.
http://www.nla.gov.au/david-branagan/in-great-haste-the-
knight-in-the-old-brown-hat

[10] Australian War Memorial. Habeita. Carinya Camp.
http://cas.awm.gov.au/item/C00080

[11] Wikipedia. http://en.wikipedia.org/wiki/Tall_al_Kabir

[12] Australian War Memorial. 51st Battalion.
http://www.awm.gov.au/units/unit_11238.asp

[13] Australian War Memorial. Somme Offensive.
http://www.awm.gov.au/units/event_158.asp

[14] Australians on the Western Front. Mouquet Farm 8 August to 3
September 1916.
http://www.ww1westernfront.gov.au/battlefields/mouquet-
farm-1916.html

[15] Australians at War. The Battle for Mouquet Farm.
http://www.australiansatwar.gov.au/stories/stories_ID=181_
war=W1.html

[16] Australian War Memorial. Mouquet Farm.
http://cas.awm.gov.au/item/J00181

[17] Australians on the Western Front. Mouquet Farm AIF Memorial.
http://www.ww1westernfront.gov.au/mouquet-farm/index.html

[18] Australian War Memorial. Mouquet Farm.
http://cas.awm.gov.au/item/H15927

[19]The AIF Project. Alexander James Stevens.
http://www.aif.adfa.edu.au/showPerson?pid=287616

Recommended Reading

- Nuttall, J.H. The War Diary of the 51st Battalion A.I.F. 1916-1919. Written and compiled by Maj. J. H. Nuttall, Unit Historian. http://www.army.gov.au/51fnqr/History.asp.
- Bean, C.E.W. (1941). Official History of Australia in the War of 1914-1918. Volume III. The Australian Imperial Force in France, 1916 (12th edition). Chapter XXI. The Advance to Mouquet Farm. Pages 726 to 770. http://www.awm.gov.au/histories/first_world_war/volume.asp?levelID=67889
- Bennett, Scott (2011). Pozières: The Anzac Story. Scribe Publications, ISBN 978-1921640353.

Chapter 5: Return to Australia

Killed in Action?

Jack's name appeared in the list of men Killed in Action as reported by the Argus (Melbourne) of Tuesday 22nd August 1916[1]:

AUSTRALIAN CASUALTIES. 197th LIST. 233 DEATHS.
The outstanding feature in the 197th casualty list is that three officers and 171 men have been killed in action The dates indicate that, while the Victorians fell mainly in the first fighting about Pozières, many New South Welshmen, Queenslanders, and Tasmanians were killed in the "brilliant success" of the 6th inst, when the German line was pierced on a front of 3,000 yards. One officer and 57 men died of wounds, and 44 officers and 569 men comprised the wounded. Details are as follow:

KILLED IN ACTION (WESTERN AUSTRALIA)
Sgt, T. Stcdman, L. Cnls. W. A. Walker, J. Dean, S. W. Dennis, Ptes. ***J. J. Griffin***, *W. J. Smith, J. Honey, C. F. Shepherd, M. Solomon, J. T. Lee, D. J. McArthur,K. J. Ryan, W. C. Hollings, A. K. Gillett, R. W. Jamieson, S. Lacey, A. J. Mitchell, A. W. Copley.*

As Jack was a widower his service record gives his sister Madge as his next of kin. Madge must have been alarmed to see her brother's name in the list of men killed in action. However, his name was listed in error and the following month the army sent a telegram to Madge reporting that her brother had been wounded in action and a further telegram dated the 22nd September 1916 informed her that he had received a gunshot wound and had been admitted to the Second Eastern General Hospital (see pages 65).

AUSTRALIAN CASUALTIES.

197th LIST.

233 DEATHS.

The outstanding feature in the 197th casualty list is that three officers and 171 men have been killed in action. The dates indicate that, while the Victorians fell mainly in the first fighting about Pozieres, many New South Welshmen, Queenslanders, and Tasmanians were killed in the "brilliant success" of the 6th inst., when the German line was pierced on a front of 3,000 yards. One officer and 57 men died of wounds, and 44 officers and 569 men comprised the wounded.

Details are as follow:—

KILLED IN ACTION.

WESTERN AUSTRALIA.—Sgt. T. Stedman, L.-Cpls. W. A. Walker, J. Dean, S. W. Dennis, Ptes. J. J. Griffin, W. J. Smith, G. Honey, C. F. Shepherd, M. Solomon, J. T. Lee, D. J. McArthur, F. J. Ryan, W. C. Hullings, A. K. Gillett, R. W. Jameson, S. Lacey, A. J. Mitchell, A. W. Copley.

TASMANIA.—Sgt. A. G. Tolman, L.-Cpls. H. C. Nicholas, W. Ludbey, Ptes. D. Barclay, E. J. Howard, H. A. Burrell, W. Smith.

The Argus (Melbourne, Tuesday 22nd August 1916, page 5 mistakenly reports Jack as KIA.

MRS. E. JONES

CARE RZY WALKINS XXXX LIMITED

YORK STREET

ALBANY (W.A)

REGRET REPORTED BROTHER PRIVATE JOHN J.GRIFFIN WOUNDED

WILL PROMPTLY ADVISE IF ANYTHING FURTHER RECEIVED.

BASE RECORDS

14/9/16

Telegram sent to Jack's sister, Madge, on the 14th September 1916.

MRS. E. JONES

CARE XXXX RZY WALKINS LIMITED

YORK STREET

ALBANY (W.A)

NOW REPORTED BROTHER JOHN J.GRIFFIN ADMITTED 2nd

EASTERN GENERAL HOSPITAL 15th AUGUST GUNSHOT WOUND ARM MILD WILL

PROMPTLY ADVISE IF ANYTHING FURTHER RECEIVED.

BASE RECORDS

22/9/16

Telegram sent to Jack's sister, Madge, on the 22nd September 1916.

Wounded in Action

Jack's service record shows that he was wounded in action on the 14th August 1916, and subsequently evacuated to England on *HS Antwerpen* from Boulogne on 21st August 1916 (see pages 56, 67)[2]. Jack had received a gunshot wound and had been admitted to the Second Eastern General Hospital (see page 67). This hospital was located in Brighton, England[3]. The Second Eastern General Hospital at Brighton occupied a boys' grammar school and several elementary schools. During the war, ambulance trains carried 30,070 patients to Brighton. The 2nd Eastern contained 98 officer beds and 1190 other ranks beds[3].

On the 1st November 1916 Jack was moved to the First Auxiliary Hospital and on the 9th November he was transferred to the AIF Command Depot at Weymouth (see page 67)[2]. Weymouth was where injured ANZAC soldiers were sent to convalesce following their discharge from hospital[4]. A medical report on

The hospital ship HS Stad Antwerpen. She is shown here during World War I, the photograph was probably taken at Dover in 1915[5].

Statement of service showing that after being WIA, Jack was evacuated to England on the hospital ship HS Stad Antwerpen at Boulogne 21st August 1916. He was admitted to the Second Eastern General Hospital at Brighton on 22nd August with a "G. W. Arm" (gunshot wound arm). Returned to Australia 13th February 1917 on HMAT Benalla from Plymouth "for change 5 mths G. S. W. R. up arm and F. arm" (gunshot wound right upper arm and forearm).

Casualty Form-Active Service 2. Service record showing Jack's transfer from the Second Eastern General Hospital in Brighton to the First Auxiliary Hospital on the 1st November 1916. Also shown is his subsequent transfer to Weymouth on the 9th November. He returned to Australia on the 13th February 1917 aboard HMAT Benalla and was discharged from the army on medical grounds on 11th June 1917.

10th November (see page 69) records: "Moderate flexion of 3rd and 4th fingers of right hand - slight amount of flexion of 2nd - none of 1st - no flexion of thumb. Hand at present useless."

Madge received several telegrams (see pages 65) and a letter (see page 69) from the army reporting that Jack had been wounded in action and keeping her informed of his progress. He was finally sent home to Australia on 13th February 1917 on *HMAT Benalla* from Plymouth "for change 5 mths G. S. W. R. up arm and F. arm" (gunshot wound right upper arm and forearm), see page 67.

No. 2 Australian Command Depot Weymouth

Weymouth was the depot for the ANZAC casualties sent to UK hospitals for treatment and then discharged as convalescent[4]. In 1916 the AIF Command Depot at Weymouth accommodated those men not expected to be fit for duty within six months. Over 120,000 Australian and New Zealand troops passed through Weymouth during World War I.

HMAT Benalla, the troopship that brought Jack back to Australia in 1917. The Benalla weighed 11,118 tons with an average cruise speed of 14 knots or 25.92 kph. She was owned by the P & O SN Co, London, and leased by the Commonwealth until 6th August 1917[6].

Medical report, Weymouth, 10th November 1916. Jack was declared unfit for service following injuries received in battle. The hand written note reads: "Moderate flexion of 3rd and 4th fingers of right hand - slight amount of flexion of 2nd - none of 1st - no flexion of thumb. Hand at present useless".

Letter sent to Jack's sister, Madge on the 17th March 1917 informing her of Jack's return to Australia.

John Joseph Griffin. Postcard sent by Jack in 1916 from England to his sister in Western Australia. Printed on the postcard is: "By photographer Walter J. Dovey, 13 St. Thomas St., Weymouth." On the back of the postcard, handwritten in ink: "John Joseph Griffin, b. 1880 [sic] Killorglin Co. Kerry. d. 1926 Roebourne W. Aust. 45 yrs." Also faintly in pencil, in different handwriting, possibly Jack's is: "51st Batt".

The ANZAC Memorial on the north end of Weymouth Esplanade commemorates the Australian and New Zealand troops who were accommodated in camps in the town during World War I[4].

Photograph

While recuperating in Weymouth, Jack went to the premises of professional photographer Walter J. Dovey at 13 St. Thomas St., Weymouth to have his photograph taken. The picture was printed in the form of a postcard and sent in an envelope to his sister Madge in Western Australia. No doubt Madge was pleased to receive positive proof that her brother was still alive.

ANZAC Memorial, the Esplanade, Weymouth.

The Final Days

Jack received three medals for his military service: the 1914-15 Star, the British War Medal and the Victory Medal[2]. The combination of a Star, Victory Medal and War Medal was often referred to by the nickname, "Pip, Squeak and Wilfred" after the popular cartoon characters from the Daily Mirror of that period[7].

After Jack returned to Australia he was treated for his injuries at No. 8 Australia General Hospital in Fremantle (see Appendix I). He was discharged from the army on medical grounds on 11th June 1917 and allotted an army pension of fifteen shillings a fortnight commencing on the 12th June 1917. According to family recollections Jack was too proud to accept his pension preferring to return to work at Mulga Downs not far from Maroonah Station in Western Australia[8]. His position at Mulga Downs was now one of Station Overseer.

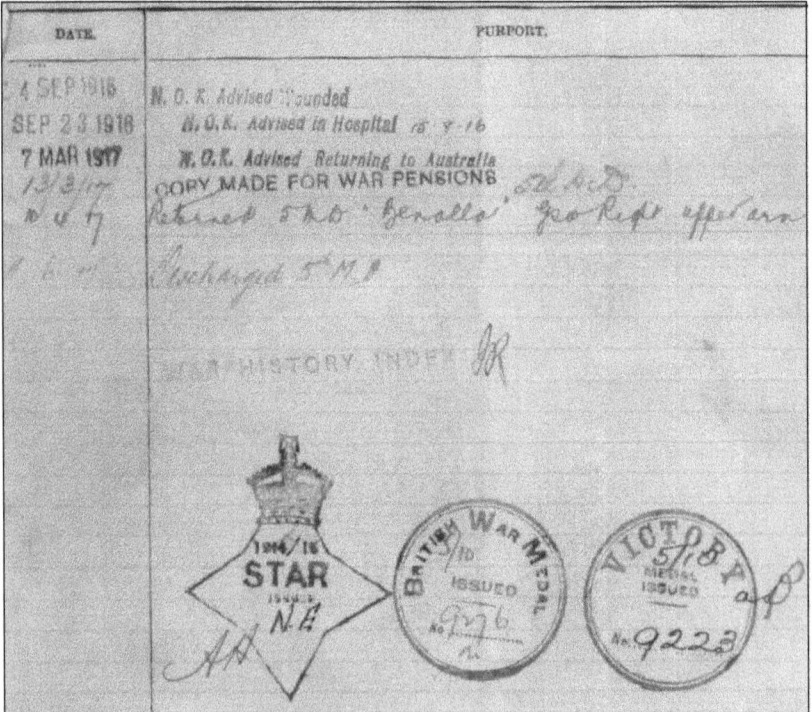

Medals received. The final page of Jack's service record.

Details of Jack's army pension.

The Old Roebourne Cemetery, Western Australia, where Jack is buried.

DEATH IN THE STATE OF WESTERN AUSTRALIA

ROEBOURNE.

Deaths in the .... District of Western Australia, Registered by ....

Jack's death certificate showing that he died of supperative pancreatitis, pancreatic abscesses, heart failure.

He became ill in April 1926 and died in Roebourne Hospital on the 20th of May 1926 following a month's illness. The cause of death is given as supperative pancreatitis, pancreatic abscesses, and heart failure. He was buried the following day at Roebourne Cemetery (Old Roebourne Cemetery); the officiating minister was H. H. Simpson, an Anglican Archdeacon. His grave reads "In Loving Memory of JOHN JOSEPH GRIFFIN late AIF who died 21st May 1926 aged 45 years."

On the 25th of April every year (ANZAC day) Australians remember those who fought with the AIF during World War I. The acronym ANZAC stands for Australian and New Zealand Army Corps and Anzac Day is one of the most important national occasions in both Australia and New Zealand.

Jack is remembered by his relatives in Australia as "a quick tempered man but he was also kind and generous and lavished gifts on his younger sister Madge and her children"[8].

His final resting place.

References

[1] The Argus (Melbourne) of Tuesday 22nd August 1916

[2] National Archives of Australia. Search for "Griffin 4797"at
http://naa12.naa.gov.au/

[3] The 2nd Eastern General Hospital, Brighton.
http://www.wartimememoriesproject.com/greatwar/hospitals/2
easterngeneralhospital.php

[4] The ANZAC Memorial.
http://memorials.dva.gov.au/MemorialDetail.aspx?Id=105

[5] Stad Antwerpen.
http://users.telenet.be/eddy.lannoo/images/151/storyline/1913
StadAntwerpen.htm

[6] His Majesty's Australian Transports.
http://alh-research.tripod.com/ships_lh.htm
http://memorials.dva.gov.au/MemorialDetail.aspx?Id=105

[7] Pip, Squeak and Wilfred (Medals).
http://www.firstworldwar.com/atoz/pipsqueakwilfred.htm

[8] Jones, A.E. (1998). Mansfield of Maroonah: From West Ireland
to the Ashburton District of North West Australia 1874-1913
and beyond. Bibra Lake, WA. ISBN 0957788355.

Epilogue

Jack's mother, Elizabeth, died in 1902 not long after Jack emigrated to Australia. His father, Edward, passed away in 1916 while Jack was serving with the AIF. Although Jack and his wife Mary had no children, descendants of their brothers and sisters are living today in various parts of the world.

At the time of Mary's death three of her brothers, Gregory, John and James were living in Carnarvon, WA. John Glass and Gregory Glass both have descendants living in Western Australia. Mary had three other siblings, Annie, Maud and William whose descendants live in Northern Ireland and another sister, Elizabeth Malone (*née* Glass) who emigrated from Northern Ireland to America. Elizabeth's descendants currently live in various parts of the USA.

After Jack's sister Madge left Maroonah, she met a young English Methodist minister named Edward Jones. They married in 1913 and had three children, the youngest being Alwyn Edward Jones[1]. Following their marriage they lived in Perth, WA and later moved to White Road, Bunbury. Tom Naughton, Madge's son from her previous marriage, also lived with them at this time. By May 1916 the family had moved to Cliff Street, Albany, later moving to Kalgoorlie and finally settling in Swanbourne, WA.

Following the loss of Maroonah, Jack's cousin George Harman Dixon, originally from Enniscorthy in Ireland, remained in Australia. He enlisted in the AIF in July 1915 and was later promoted to sergeant. On leaving the army in 1919 he obtained a soldier settlement farm at Bridgetown, WA. George married a local girl and they had three children[1].

Jack's nephew Tom, who had been born in Ireland and emigrated to Australia as a young child with his mother, initially gained employment with a footware company and later worked for George Harman Dixon at his farm in Bridgetown. Around

1930 Tom became the part-owner of a claim in the gold fields of Larkinville[2]. He later mined asbestos in Yampi Gorge and with the proceeds from his gold and asbestos mining ventures he was able to finance the purchase of a small dairy farm in Byford. Here he met and married his wife Ellen Mary ("Marie") in 1941 and they had a daughter, Joan. During World War II Tom served with the Second Australian Imperial Force as a corporal[3]. In later years he returned to work in the footware industry[1].

Jack's brother Edward Blake "Ned" Griffin remained in Ireland and married Mary Jane Agnew. They had nine children and many grandchildren and great-grandchildren. One of Ned's sons, Edward ("Ted"), emigrated to Africa and later moved to Western Australia. Ned's eldest son, George Mansfield Griffin, remained in Ireland and became a teacher like his father and grandfather before him. He is survived by four children. George's son, Hugh, is the author of this book.

Prior to his death in 1926, Jack was working as overseer on Mulga Downs Station, at that time owned by the Hancock family. Many years later, on the 16th of November 1952, Lang Hancock discovered the world's largest deposit of iron ore in the nearby Pilbara region and became one of the richest men in Australia[4]. According to *Forbes Asia* his daughter Gina Rinehart is now (2011) Australia's richest person with a net worth of A$9bn[4].

References

[1] Jones, A.E. (1998). Mansfield of Maroonah: From West Ireland to the Ashburton District of North West Australia 1874-1913 and beyond. Bibra Lake, WA. ISBN 0957788355.

[2] Bridge, Peter J. (1999). The Eagle's Nest. Larkinville, the Golden Eagle, and the Great Depression. Pages 80-83 and page 199. Hesperian Press, Carlisle, WA. ISBN 0859052745.

[3] National Archives of Australia. Search for "Naughton WX27149"at http://naa12.naa.gov.au/

[4] Wikipedia http://en.wikipedia.org/wiki/Lang_Hancock

Appendix I. Service Record

The following is a complete copy of the AIF service record of John Joseph Griffin, rank of private, serial number 4797. Jack was one of the 3,000 Irish-born Australians that volunteered for service during World War I. He enlisted at Blackboy Hill, Western Australia in December 1915 and he was discharged from the army on medical grounds in June 1917.

During his time in the army he served in Australia, Egypt and France. He fought in the Battle of Mouquet Farm, part of the Somme Offensive and received a gunshot wound to the right arm. He was awarded three medals and a army pension.

A 14720.

TRANSFERRED TO *51st Battalion*

16th Reinf. 51st Batt.

AUSTRALIAN MILITARY FORCES.

AUSTRALIAN IMPERIAL FORCE.

Attestation Paper of Persons Enlisted for Service Abroad.

No. 4747   Name *GRIFFIN JOHN JOSEPH*

Unit *9 Depot*

Joined on *29-12-15*

Questions to be put to the Person Enlisting before Attestation.

1. What is your Name?	1.	*John Joseph Griffin*
2. In or near what Parish or Town were you born?	2. In the Parish of	in or near the Town of *Kilarney* in the County of *Ireland.*
3. Are you a natural born British Subject or a Naturalized British Subject? (N.B.—If the latter, papers to be shown.)	3.	*N. B.*
4. What is your age?	4.	*34 yrs 10 mths*
5. What is your trade or calling?	5.	*Drover*
6. Are you, or have you been, an Apprentice? If so, where, to whom, and for what period?	6.	*3 yrs. R. Hillard Ironmongers.* *New St, Kilarney, Ireland.*
7. Are you married?	7.	*No.*
8. Who is your next of kin? (Address to be stated)	8.	*Sister Mrs E. Jones White Rd Bunbury W A*
9. Have you ever been convicted by the Civil Power?	9.	*No.*
10. Have you ever been discharged from any part of His Majesty's Forces, with Ignominy, or as Incorrigible and Worthless, or on account of Conviction of Felony, or of a Sentence of Penal Servitude, or have you been dismissed with Disgrace from the Navy?	10.	*No.*
11. Do you now belong to, or have you ever served in, His Majesty's Army, the Marines, the Militia, the Militia Reserve, the Territorial Force, Royal Navy, or Colonial Forces? If so, state which, and if not now serving, state cause of discharge	11.	*No*
12. Have you stated the whole, if any, of your previous service?	12.	*Yes.*
13. Have you ever been rejected as unfit for His Majesty's Service? If so, on what grounds?	13.	*No*
14. (For married men, widowers with children, and soldiers who are the sole support of widowed mother)— Do you understand that no separation allowance will be issued in respect of your service beyond an amount which together with pay would reach eight shillings per day?	14.	
15. Are you prepared to undergo inoculation against small pox and enteric fever?	15.	*Yes.*

3. *John Joseph Griffin* do solemnly declare that the above answers made by me to the above questions are true, and I am willing and hereby voluntarily agree to serve in the Military Forces of the Commonwealth of Australia within or beyond the limits of the Commonwealth.

And I further agree to allot not less than ~~two-fifths~~ three-fifths of the pay payable to me from time to time during my service for the support of my ~~wife.*†~~ wife and children.

Date *29-12-15*

Signature of person enlisted.

*This clause should be struck out in the case of unmarried men or widowers without children under 18 years of age.
†Two-fifths must be allotted to the wife, and if there are children three-fifths must be allotted.

CERTIFIED TRUE COPY

LIEUT.

AUSTRALIAN 　 **MILITARY FORCES.**

AUSTRALIAN IMPERIAL FORCE.

Attestation Paper of Persons Enlisted for Service Abroad.

No. 4797 　 Name GRIFFIN JOHN JOSEPH

Unit 39 Depot 15th Batt.

Joined on 29/12/15

Questions to be put to the Person Enlisting before Attestation.

1. What is your Name?		GRIFFIN, JOHN JOSEPH.
2. In or near what Parish or Town were you born?		2. In the Parish of .... in or near the Town of Kilarney in the County of Ireland
3. Are you a natural born British Subject or a Naturalised British Subject? (N.B.—If the latter, papers to be shown.)		Natural Born British subject
4. What is your age?		34 Yrs 10 mos
5. What is your Trade or Calling?		Driver
6. Are you, or have you been an Apprentice? If so, where, to whom, and for what period?		3 yrs R. Villard Ironmonger New St Kilarney Ireland
7. Are you married?		No
8. Who is your next of kin? (Address to be stated)		Sister, Mrs E Jones White Road Bunbury W.A.
9. Have you ever been convicted by the Civil Power?		No.
10. Have you ever been discharged from any part of His Majesty's Forces with Ignominy, or as Incorrigible and Worthless, or on account of Conviction of Felony, or of a Sentence of Penal Servitude, or have you been dismissed with Disgrace from the Navy?		No.
11. Do you now belong to, or have you ever served in, His Majesty's Army, the Marines, the Militia, the Militia Reserve, the Territorial Force, Royal Navy, or Colonial Forces? If so, state which, and if not now serving, state cause of discharge?		No
12. Have you stated the whole, if any, of your previous Service?		Yes
13. Have you ever been rejected as unfit for His Majesty's Service? If so, on what grounds?		No
14. (For married men and widowers with children and soldiers who are the sole support of widowed mother)— Do you understand that no Separation Allowance will be issued in respect of your service beyond an amount which, together with pay, would reach eight shillings per day ...		
15. Are you prepared to undergo inoculation against smallpox and enteric fever?		Yes

I, John Joseph Griffin do solemnly declare that the above answers made by me to the above questions are true, and that I am willing to serve in the Military Forces of the Commonwealth of Australia within or beyond the limits of the Commonwealth.

And I further agree to allot not less than *two-fifths* three-fifths of the pay payable to me from time to time during my service for the support of my wife and children.

Date 4th December 1915 　　　 John J. Griffin

Signature of Person Enlisted.

* This clause should be struck out in the case of unmarried men or widowers without children under 18 years of age.
† Two-fifths must be allotted to the wife, and if there are children three-fifths must be allotted

2

CERTIFICATE OF ATTESTING OFFICER.

The foregoing questions were read to the person enlisted in my presence.

I have taken care that he understands each question, and that his answer to each question has been duly entered as replied to by him.

I have examined his naturalization papers and am of opinion that they are correct.

(This to be struck out except in the case of persons who are naturalized British Subjects.)

Date _30-12-15_

Signature of Attesting Officer.

Lieut

OATH TO BE TAKEN BY PERSON BEING ENLISTED.*

3, _John Joseph Griffin_ swear that I will well and truly serve our Sovereign Lord the King in the Australian Imperial Force from _30-12-15_ until the end of the War, and a further period of four months thereafter unless sooner lawfully discharged, dismissed, or removed therefrom; and that I will resist His Majesty's enemies and cause His Majesty's peace to be kept and maintained; and that I will in all matters appertaining to my service, faithfully discharge my duty according to law.

So Help Me, God.

Signature of Person Enlisted.

Taken and subscribed at _Blackboy Hill_ in the State of _W. Australia_ this _30th_ day _Dec._ of _____ 19 _15_, before me—

Roebourne

Signature of Attesting Officer.

Lieut

* A person enlisting who objects to taking an oath may make an affirmation in accordance with the Third Schedule of the Act, and the above form must be amended accordingly. All amendments must be initialed by the Attesting Officer.

3

Description of *John Joseph Griffin* on Enlistment.

Age 34 years 10 months.	DISTINCTIVE MARKS.
Height 5 feet 8 inches.	
Weight 15.0 lbs.	*nil*
Chest Measurement 46 inches.	
Complexion *fresh*	
Eyes L. *Blue*	
Hair D. *brown*	
Religious Denomination *C.E.*	

CERTIFICATE OF MEDICAL EXAMINATION.

I HAVE examined the above-named person, and find that he does not present any of the following conditions, viz.:—

Scrofula; phthisis; syphilis; impaired constitution; defective intelligence; defects of vision, voice, or hearing; hernia; hæmorrhoids; varicose veins, beyond a limited extent; marked varicocele with unusually pendent testicle; inveterate cutaneous disease; chronic ulcers; traces of corporal punishment, or evidence of having been marked with the letters D. or B.C.; contracted or deformed chest; abnormal curvature of spine; or any other disease or physical defect calculated to unfit him for the duties of a soldier.

He can see the required distance with either eye; his heart and lungs are healthy; he has the free use of his joints and limbs; and he declares he is not subject to fits of any description.

I consider him fit for active service.

Date 4 Dec 15

Place Rozbourne

I. Maunsell
Signature of Examining Medical Officer.

CERTIFICATE OF COMMANDING OFFICER.

I CERTIFY that this Attestation of the above-named person is correct, and that the required forms have been complied with. I accordingly approve, and appoint him

to ....

Date 28 FEB 1916

Lt. Col.,
Camp Commandant

Place Blackboy Hill   Commanding ....

84

Org Base ?
13.2.17

4

Statement of Service of No. 4797   Name   Griffin J.J.

Unit in which served.	Promotions, Reductions, Casualties, &c.	Period of service in each rank. From—	To—	Remarks.
39 Dep. C Company 5th Dep Battalion 15/n Reinforcements	Private	29-12-15		816 6 5 0
	TRANSFERRED TO 51st Battalion 20/5/16			E 4861 29/5/16
	Disch'd from No.26 Gen Hosp. W Base. Etaples.	1.7.16		v 19/2798 12/7/16
	Disc from No.26 Gen Hosp & joined Base Details	8.7.16		n 21/3021 22/7/16
	TAKEN ON STRENGTH FROM B/6.	22-7-16		24-3341 7/8/16
	WOUNDED in Action France between	19/16-8-16		24-3804 29.8.16
	Embarked for England on HS Western at Boulogne SARB Qn	21.8.16		842. 30.3991. 3-9-16
	Adm. 1 Eastern Gen Hosp Brighton (S.W. Arm)	22-8-16		P-11-28/535F 6/10/16
12.2.17 adm Hd qrts Clew Roll 51st Bn	Retd to Aust per H.T Benalla for change 3 mths G.S.W.R. up: arm & face. Plymouth	London	13-2-17	LR 165 D/o 191 E 9.3.17 9.3.17 13.2.17 adm H dqrs Now Roll

I have examined the above details, and find them correct in every respect.

Statement of Service of No. 4797 Name Griffin J. J.

Unit in which served.	Promotions, Reductions, Casualties, etc.	Period of Service in each Rank. From—	To—	Remarks.
39 Dep.	Private.	29-12-15		8/6 650
6 Company 5th Dep Bat	Discharged from No 8 N.M. at Fremantle to duty	16.3.16		6/6, No 8. 434 DHR 75
13/11 Rank				
51 Batt	Private	20/5/16		BR 61/50 BR 62/8305
11th Bn	Returned to Penalta 7th M.D. Discharged 8th M.D.		26 17	B.R. 191/306

I have examined the above details, and find them correct in every respect.

E. Simmons Lt

1270—15-6.

Army Form B. 103.

Casualty Form — Active Service.

Regiment or Corps _"C" COMPANY_ _3rd TRAINING BATTALION_

Rank _Pte._ Name _Giuseppe Montague Joseph._

Regimental No. _4797_

Enlisted (a)

Term of Service (a)

Service reckons from (a)

Date of promotion to present rank } appointment to lance rank }

Numerical position on roll of N.C.O.'s. }

Extended

Re-engaged

Qualification (b)

REPORT		Record of promotions, reductions, transfers, casualties, &c., during active service, as reported on Army Form B 213, Army Form A. 36, or in other official documents. The authority to be quoted in each case.	Place	Date	Remarks taken from Army Form B. 213, Army Form A. 36 or other official documents.
Date	From whom received				
	C.O. 3TB 3rd Bn.	Died at	Suez Alexandria	25.4.18	A P 7372 ex Ulysses D7517
	C.O. HT Huntspill	Embarked	Alexandria	7.6.16	A P 8298
		Disembarked	Marseilles	14.6.16	L R 5811
3716 No Gen. 5.7.16 67.16			Etaples	6.7.16 6.7.16 6.7.16	at 7372 at 178 at 7372
29-7-16 C.O.	Taken on strength				
	51 Bn.213	ex 4 Div, Base Dpt.	France	22-7-16	A.Q.1716 D.O.
		C.O Bn. Wounded in action 14-8-16			
		In Field AQ3263			

(a) In the case of a man who has re-engaged for, or enlisted into Section D, Army Reserve, particulars of such re-engagement or enlistment will be entered.

(b) Signaller, Shoeing Smith, &c., &c., also special qualifications in technical Corps duties.

(P.T.O)

REPORT		Record of promotions, reductions, transfers, casualties, &c., during active service, as reported on Army Form B. 213 Army Form A. 36, or other official documents. The authority to be quoted in each case.	Place	Date	Remarks taken from Army Form B. 213, Army Form A. 36, or other official documents.
Date	From whom received				
19.8.16	Hamilton H.Qr. o M.N.Army		Boulogne	19.8.16	
7.8.16					
28.8.16.	Captain	S.W. Ann.	England.	28.8.16	28/53 E.R. 1657
	Q. M.O. Brighton				
4.11.16.	9/3 r4.	Taftd to 1st Aux Hosp from 2nd East G Hosp (S.W.Fome R. Appl.)	England.	1.11.16	IR 443.
14.11.6.	JB 80.	Disd to Weymouth fr 1st Aux Hosp.	England	9.11.6	IR 443
10.11.6		M.I. from Harefield	Weymo. C.	9-11-16	EB 8029.
13.2.17.	Adm Hd Qrs Mons R.Bt.	Returned to Australia per. H.T. "Benalla" for change S. mth. G.S.N. R. 14 Arm's Ham ex Plymouth	London	13.2.17.	LR165 D/14/E.9.2.17. 157
		Discharged 5th MD		11.6.17	B. Oct. 94 Page 50.
					No. 1. 361 14.5.16

LCB

BASE RECORDS OFFICE, A.I.F.,
12th May, 1916.
Victoria Barracks,

MELBOURNE,

Dear Madam,

The your change of address on the records of your dated 28th ulto

Soldier No. 4797 Private J. J. Griffin, 15th Reinforcements, 11th Battalion,

is hereby acknowledged and will receive attention.

Yours faithfully,

J.M. LEAN, Major.
xxxxx
Capt.
Officer i/c Base Records.

DESPATCHED

In all communications reference
number, rank, full name and unit
of Soldier referred to, is to be stated.

Mrs. E. Jones,
C/o Maywalkins Ltd.,
York Street,
ALBANY, W.A.

Albany May 28

To Base Records Office
Defence Department
Melbourne

Dear Sir

Some weeks ago I wrote re
change address to "Military Head quarters
Expeditionary Force Melbourne" but since
learned that I should have addressed
my letter to above. If by chance my
letter reached you, you have this information.
If not I hereby inform you that being
next of kin to Pte J J Guppis 4797.
15711th Battalion. A 97. I have changed
my address from White Rd Bunbury
W. A., to C/o Ezywalkins Ltd York St.
Albany W. A

Yrs faithfully

M E Jones
(Mrs E Jones) sister)

90

Mrs. R. JONES

CARE RZY WALKING XXXX LIMITED

YORK STREET

ALBANY (W.A)

REGRET RECORDS REPORTED BROTHER PRIVATE JOHN J.O.GRIFFIN WOUNDED

WILT PROMPTLY ADVISE IF ANYTHING FURTHER RECEIVED.

BASE RECORDS

14/9/16

MRS. E. JOHNS

CARE XXXX EZY WALKINS LIMITED

YORK STREET

ALBANY (W.A)

NOW REPORTED BROTHER JOHN J. GRIFFITH ADMITTED 2nd

EASTERN GENERAL HOSPITAL 15th AUGUST GUNSHOT WOUND ARM MILD WILL

PROMPTLY ADVISE IF ANYTHING FURTHER RECEIVED.

BASE RECORDS

22/9/16

COMMONWEALTH OF AUSTRALIA.   Form K.

The War Pensions Act 1914.

MEDICAL CERTIFICATE.

P. M. O'Meara Major.   P. M. R. B.

I, M.S. C. Moore.   Major.   5th Mil. Dist.   hereby declare

that I have this day examined Griffin J. J.

of Court Hotel, Beaufort St. Perth.
a claimant for a pension under the above-named Act.

I find that the claimant—

(1) Is about 27 years of age.

(a) Fully describe claimant's general condition.

(2) Is suffering from (a) G. S. Wound right arm & fore arm

Wounds healed Scars adherent. Muscular power weakens

expecially right index finger. Desides discharge.

(b) Give information showing whether the condition has resulted from claimant's employment in connexion with warlike operations.

(3) The above condition is the result of (b) G. S. Wounds in France.

(c) Fill in period of time.

(d) Insert "due" or "not due," as the case requires.

(4) It has in my opinion existed for (c) nine months

and is (d) not due to the default of the claimant.

(5) The condition is such as to render claimant incapacitated for

(e) State period of time.

work for the period of (e) 12 months from this date.

(f) If earning power wholly lost state "the whole." If only partially lost give the fraction which has been lost, as, for example, "one-half" or "three-fourths."

(6) The claimant has lost his earning power to the extent of (f)

One quarter

(Sgd) P. M. O'Meara. Major.

" S. C. Moore Major.   P. M. R. B.

Commonwealth Medical Referee.

(Address) No. 8. A.G.H. FREMANTLE.

Date 10th May. 1917.   191

To the

DEPUTY COMMISSIONER OF PENSIONS,

Perth.

D.5366 16—C.3372—500.

Upper Berth. ARMY FORM B. 178.

To be used (a) for recruits enlisting direct into the Regular Army, and (b) for men of the Territorial Force when they are admitted to Hospital. Army Form B. 178ᴬ to be used for Special Reserve recruits and Special Reservists enlisting into the Regular Army.

MEDICAL HISTORY of

4366

Surname _GRIFFIN_   Christian Name _J. J._

TABLE I.—GENERAL TABLE.

Birthplace ... Parish _____ County _Ireland_

Examined ... . { on _____ day of _____ 191 , { at _____

Declared Age ... ... _35_ years _____ days.

Trade or Occupation ... _Italian Overseer._

Height ... ... _____ feet _____ inches.

Weight ... ... _____ lbs.

Chest Measurement { Girth when fully Expanded _____ inches. { Range of Expansion _____ inches.

Physical Development ...

	Right	Left
Vaccination Marks { Arm ...		
{ Number		

When Vaccinated ... ...

Vision ... ... { R.E.—V= { L.E.—V=

(a) Marks indicating congenital peculiarities or previous disease ... ... (a)

(b) Slight defects but not sufficient to cause rejection ... ... (b)

Approved by (Signature) _____
(Rank) _____
Medical Officer.

Enlisted ... ... { at _Perth_ { on _29_ day of _December_ 191 _5_.

Joined on Enlistment ...	Corps.	Regtl. No.
	51 Btn.	4797
Transferred to		

Became non-effective by ...

_____ on _____ day of _____ 191 .

(Signature) _____
(Rank) _____

(1116) 200M 1/16 N.P.A. Ltd.   P.T.O.

94

Table II.—Only for Admissions to Hospital or to the Si

Name of Hospital.	Admitted to Hospital.			Discharged from Hospital.			Disease.	Number of Days in Hospital.
	Day.	Month.	Year.	Day.	Month.	Year.		
	1	11	16	9	11	16	S.W. R.fmur + an.	8
Hm. J. Benalla	11	2	17					
No 8 a E.H. Finkle	3	4	17	10	5	17	E. S.W. Rt arm	

Part 1

Upper Berth.

List in the case of Warrant Officers treated in quarters.

Remarks bearing on the case, nature, or treatment of the case, likely to be of interest or of future use. In cases of syphilis, admissions and re-admissions to hospital will be shown. The subsequent progress, including particulars of treatment out of hospital, transfers, &c., will be given in the special syphilis case sheet.	Signature of Medical Officer.
SW. R. forearm & arm Aug 16th some loss of power in hand + sensation Muscles both of forearm supplied by MS respond to Farad. also muscles supplied by U. nerve , Flex. C. Rad., Palm. Longus, Flex. Profundus, + ² Pronator Teres Remainder of muscles supplied by M nerve do not respond to Farad. Inj y of M nerve (C S O)	H M William
nil to report	Mary Melland Mason
Trans to Details Camp DPU	W M Coton Captn

Part 2

Table III. Boards; Courts of Inquiry, Vaccination, Inoculations, etc.; Examinations for Field or Foreign Service, Extension, Re-engagement, or Prolongation of Service; Issue of Surgical Appliances; Particulars of Dental Treatment, etc.

Date.	Brief details, and signature.
	BOARDED 10th November 1916
	FINDING:- Permanently unfit for general service for
	more than 6 months and unfit for home service
	[signature] Major.
	S.M.O. No. 2 AUSTN. COMMAND DEPOT.

[Stamp: No. 2 AUSTL. COMMAND DEPOT A.M.C. 10 NOV 1916 No. ...... WEYMOUTH]

Table IV.—Service Table.

Station or Troopship.	Date of arrival or embarkation.	Date of departure or disembarkation.	Station or Troopship.	Date of arrival or embarkation.	Date of departure or disembarkation.

Army Form B. 122

No. 47297 Name Griffin J.J. Sqn., Battry., or Company Corps 51st Bttn Date of enlistment Service or Efficiency Pay

Date of last entry in Company Conduct Sheet | Date of Discharge | pass | Period not reckoning towards freedom from extra fine | Sheet No. Signature O.C. Company, etc. G. Ashe Lieut O.C. (initial) Chatmade:

Place	Date of offence	Rank	Cause of Discharge	Offence	Names of Witnesses	Punishment awarded	Date of award or of order dispensing with trial	By whom awarded	Remarks
				Substitute 9/2/17					
Y5985.H Bramshott	26.4.17	Pte		A.O.L. from 9 a.m 25.4.17 till 10 p.m next day (1 x 5°)	Lt Metcalfe 2 other s.s	7 days pay forfeit			A.C.L. W.V.O.A.I.Hall 1.4.1
N5 985.H Bramshott	10.5.17	Pte		Entitled to further entry					
Dieren France	11.6.17			Conduct not being					C.E. Denis

W. 16619/M 927. 8,000m. J. P. & Co. Ltd. (O 1668). Forms/B.16m.

Place	Date of offence	Rank	Cause of Discharge	Offence	Names of Witnesses	Punishment awarded	Date of award or of order dispensing with trial	By whom awarded	Remarks

A.M. Book 8.

AUSTRALIAN   MILITARY FORCES.

MEDICAL CERTIFICATE.

(To accompany a Man Transferred from one Hospital to another.)

Extract from Admission and Discharge Book of _8 Aust. General H_ Hospital at _Fremantle_ Date _10/5/17_

No. of Case.	Regiment or Corps.	Squadron, Battery, or Company.	Regt. No.	RANK AND NAME. Surname first. If Married, write "M" under name.	Age last Birthday.	Service.	Service in the Command.	DATES. Admitted into Hospital.	Transferred.	Religion.	DISEASE. (a) Primary. (b) Secondary. (c) Operations.	Destination on Transfer, and to what Hospital or Ship Transferred.
132	51st Batt		6797	Pte Griffen M				3 / 4 / 17	10 / 5 / 17		GSW Rt hand &c &c arm Camp 10pu	

State here briefly reasons for Transfer, and note any particulars of Case for information of Medical Officer.

signature
Medical Officer in Charge. 8. A.G.H.

Captain

COMMONWEALTH MILITARY FORCES.

5th MILITARY DISTRICT

Telephone:

No.8.Australian General Hospital.

FREMANTLE.

May 6 1917.

J.J. Griffin   Rank. Pte

Reg No 4797   Unit 51

Request my Discharge from the A.I.F. I understand that the

issue of my Discharge will relieve the Defence Department from

any further liability excepting my Pension. I also understand

that reassessment on Pension may be considered afterwards.

J.J. Griffin .

P.Taylor   Witness.

Upper Reath

Army Form B. 179.

Medical Report on an Invalid.

4366

Station

Date

*No. 2 AUSTN. COMMAND DEPOT
A.M.C.
10 NOV 1916
WEYMOUTH*

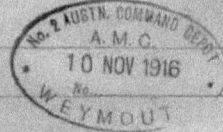

1. Unit 51st Battn. A.I.F.
2. Regimental No. 4797
3. Rank Private
4. Name GRIFFIN John Joseph

5. Age last birthday 35
6. Enlisted { on 29 December 1915 { at Perth
7. Former Trade or Occupation { Station Overseer

No 4797 Rank Pte. Name Griffin J.J. Unit 51st Batt.

I HEREBY CERTIFY that action has been taken in accordance with Circular No. 269, and the Deputy Commissioner for Pensions supplied with copy of Medical Board proceedings.

Baylor Capt
Staff Officer for Invalids,
5th Military District.

29/5/1917

§ 1584/10

11. Give concisely the essential facts of the history of the disability, noting entries on the Medical History Sheet bearing on the case.

S.W. right

Aug 14th. Injury to median nerve.

12. (a) Give your opinion as to the causation of the disability.

Shell wound

(b) If you consider it to have been caused by active service, climate, or ordinary military service, explain the specific conditions to which you attribute it (See notes on page 3).

active service

Upper Rath

Medical Report on an Invalid.

Army Form B. 179.

4366

Station ___ No. 2 AUSTN. COMMAND DEPOT A.M.C. 10 NOV 1916 WEYMOUTH

Date ___

1. Unit 51st Battn. A.I.F.
2. Regimental No. 4191
3. Rank Private
4. Name GRIFFIN John Joseph

5. Age last birthday 35
6. Enlisted { on 29 December 1915 { at Perth
7. Former Trade or Occupation { Station Overseer

8. Disability.

Gun shot wound right upper arm and forearm.

Statement of Case.

Note.—The answers to the following questions are to be filled in by the Officer in medical charge of the case. In answering them he will carefully discriminate between the man's unsupported statements and evidence recorded in his military and medical documents. He will also carefully distinguish cases entirely due to venereal disease.

9. Date of origin of disability. 14th August 1916

10. Place of origin of disability. Pozieres France

11. Give concisely the essential facts of the history of the disability, noting entries on the Medical History Sheet bearing on the case. S.W. right forearm + arm. Aug 14th. Injury to median nerve.

12. (a) Give your opinion as to the causation of the disability. Shell wound

(b) If you consider it to have been caused by active service, climate, or ordinary military service, explain the specific conditions to which you attribute it (See notes on page 2). active service

13. What is his present condition?

Weight should be given in all cases when it is likely to afford evidence of the progress of the disability.

Moderate flexion of 3rd & 4th fingers of right hand. Slight amount of flexion of 2nd. — none of 1st. Good flexion of thumb. Hand at present useless.

14. If the disability is an injury, was it caused

(a) In action? *Yes*

(b) On field service? *Yes*

(c) On duty? *Yes*

(d) Off duty? *No*

15. Was a Court of Inquiry held on the injury?

If so—(a) When?

(b) Where?

(c) Opinion?

} *not applicable*

16. Was an operation performed? If so, what? *Yes*

17. If not, was an operation advised and declined? *No*

18. *In case of loss or decay of teeth.* Is the loss of teeth the result of wounds, injury or disease, directly* attributable to active service? *not applicable*

19. Do you recommend

(a) Discharge as permanently unfit, *No*

or

(b) Change to ~~England~~? *Australia.* *yes*

_____

L t Roe Capt

Officer in medical charge of case.

I have satisfied myself of the general accuracy of this report, and concur therewith, except†

Station _____

Date _____

_____ Major.

S.M.O. No. 2 AUSTN. COMMAND DEPOT.

Officer in charge of Hospital.

* Loss of teeth on, or _____ active service, should be attributed thereto, unless there is evidence that it is due to some other cause.

† Delete this word if no exceptions are to be made.

Upper Berth.

4797
4366

21. Has th

22. Is th

23. If ne
mini

To be sig

24. To
for
gene...
present?

In defining the extent of his inability to
earn a livelihood, estimate it at ⅓, ¼, ½,
or total incapacity.

25. If an operation was advised and declined,   Not applicable
was the refusal unreasonable?

26. Do the Board recommend   Finding Temporarily unfit for general service
(a) Discharge as permanently unfit   for more than 6 months + unfit for
or   Home service
(b) Change to England?

Signatures :—   President.

Station   Members.

Date

Approved.

Station

Date

Administrative Medical Officer
D.M.S. A.I.F.

No. 9 AUSTN. COMMAND
A.M.C.
10 NOV 1916

DIRECTOR MEDICAL SERVICES
30 NOV 1916
HEAD-QUARTERS

Berth.

Note.—

in the event of ... possession of the most reliable information to enable them to ... ard, as, ... be in ...

*(ii.) Expressions such as "may," "might," "probably," &c., should be avoided.

(iii.) The rates of pension vary directly according to whether the disability is attributed to (a) active service, (b) climate, or (c) ordinary military service. It is therefore essential when assigning the cause of the disability to differentiate between them (see Articles 1162 and 1165, Pay Warrant, 1913).

(iv.) In answering question 20 the Board should be careful to discriminate between disease resulting from military conditions and disease to which the soldier would have been equally liable in civil life.

(v.) A disability is to be regarded as due to climate when it is caused by military service abroad in climates where there is a special liability to contract the disease.

20. (a) State whether the disability is the result of (i.) active service, (ii.) climate, or (iii.) ordinary military service.

active service

(b) If due to one of these causes, to what specific conditions do the Board attribute it?

gunwound.

21. Has the disability been aggravated by

(a) Intemperance ? — *No*

(b) Misconduct ? — *No*

22. Is the disability permanent ? — *No*

23. If not permanent, what is its probable minimum duration ?

To be stated in months. — *6 months*

24. To what extent is his capacity for earning a full livelihood in the general labour market lessened at present ?

In defining the extent of his inability to earn a livelihood, estimate it at ¼, ⅓, ½, or total incapacity. — *⅓*

25. If an operation was advised and declined, was the refusal unreasonable ? — *Not applicable*

26. Do the Board recommend

(a) Discharge as permanently unfit, or — *Finding. Temporarily unfit for general service for more than 6 months & unfit for home service*

(b) Change to England ?

Signatures :—

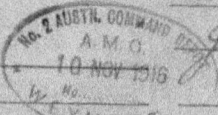

.... President.

Station ....

Date ....

Members,

Approved.

Station ....

Date ....

Administrative Medical Officer

Surgeon General,
D.M.S. A.I.F.

(On leaving Corps or Station where invalided.)

Transfer	Date			Conveyance	
or	Station		Name of	Vessel	
Embark-ation	Date			Officer in medical charge	
	Port				

Brief remarks on case during transit, and state on transfer for final disposal.

| Re-transferred | Date | |
| | Hospital or Station | |

Officer in medical charge.

(At Station or Hospital where finally disposed of.)

Station and Hospital } No 8 Australian General Hospital

Arrived from Ex Benalla   Date   3 · 4 · 17

If admitted Date	If under treatment From	To	Disease	How finally disposed of	Date of Discharge, &c.
3/4/17	3/4/17	10/5/17	G S Wounds R Arm & Fore Arm	Trans Details Camp 10/5/17	

Detailed statement as to condition on discharge and whether discharged as an invalid,
to corps, to station, or to depôt. In cases of discharge from the service it should be stated
whether the answers to questions 22, 23 and 24 are concurred in.

G. S. Wounds, right arm and fore-arm. Wounds healed.
Scars adherent. Muscular power weakened especially
right index finger. Desires discharge. Application attached.
B.F.U (22) Incapacity & (24)
for twelve months
scale

LT.-COL., O.C. No. 8 A.G.H.

Date of final Medical Board, or decision }   F. H. Oneora Major   12 in R B

10·5·7   Approved   28/5/

Administrative Medical Officer.

AUSTRALIAN IMPERIAL FORCE.

Base Records Office, A.I.F.
Victoria Barracks.

MELBOURNE.

Dear

I   to advise you that

                    has been reported

and in the event of further information coming to hand, you will
be promptly notified.

In the absence of further reports it is to be assumed
that satisfactory progress is being maintained.

It should be clearly understood that if no further advice
is received this Department has no later information to give.

Yours faithfully,

J. M. LEAN.
Major.
Officer i/c Base Records.

AUSTRALIAN IMPERIAL FORCE.

Base Records Office, A.I.F.
V   Barracks.

MELBOURNE.

Dear

I   to advise you that

                    has been reported

and in the event of further information coming to hand, you will
be promptly notified.

In the absence of further reports it is to be assumed
that satisfactory progress is being maintained.

It should be clearly understood that if no further advice
is received this Department has no later information to give.

Yours faithfully,

J. M. LEAN.
Major.
Officer i/c Base Records.

4797 Griffen J.J. 51 Btn.
Benalla (8317 P 28/7/17

WAR PENSIONS ACT 1914-1916.

Form Z.3.

DAILY STATEMENT (MILITARY)

Showing Grants, Alterations, and Cancellations of Pensions (together with date from which such action took effect); also Rejections of Claims and Deaths of Pensioners.

Statement No. 385

Containing 8 sheets

Date 11/6/17

State of **Western Australia**

1. Full name, number, rank, and unit of Member of Forces in respect of whose death or incapacity pension was claimed	Griffen, John Joseph 4797 Private 51st Battalion
2. Full name and address of person for whom pension was claimed	Griffen, John Joseph. Court Hotel, Beaufort Street Perth.
3. Relationship of such person to Member	Self
4. Result of Claim	Pension at rate of 15/- fortnightly granted from 12/6/17
5. Name and address of Trustee (if any)	

Upper Berth.

Army Form B. 179.

Medical Report on an Invalid.   4366

Station _____

Date _____

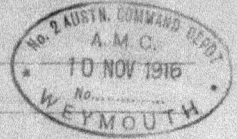

1. Unit	5 1st Batt;
2. Regimental No.	4797
3. Rank	Private
4. Name	GRIFFIN John Joseph

5. Age last birthday   35

6. Enlisted { on 29th December 1915 { at PERTH

7. Former Trade { Station Overseer
or Occupation {

8. Disability.

G.S. wound right upper arm and forearm

Statement of Case.

Note.—The answers to the following questions are to be filled in by the Officer in medical charge of the case. In answering them he will carefully discriminate between the man's unsupported statements and evidence recorded in his military and medical documents. He will also carefully distinguish cases entirely due to venereal disease.

9. Date of origin of disability.

14th August 1916

10. Place of origin of disability.

Pozieres, France

11. Give concisely the essential facts of the history of the disability, noting entries on the Medical History Sheet bearing on the case.

Shell wound right forearm and arm August 14th - Injury to median nerve

12. (a) Give your opinion as to the causation of the disability.

Shell wound

(b) If you consider it to have been caused by active service, climate, or ordinary military service, explain the specific conditions to which you attribute it (*See notes on page 3*).

Active service

13. What is his present condition?

Weight should be given in all cases when it is likely to afford evidence of the progress of the disability.

Moderate flexion of 3rd and 4th fingers of right hand - Slight amount of flexion of 2nd - None of 1st - No flexion of thumb Hand at present useless

14. If the disability is an injury, was it caused

(a) In action?   Yes

(b) On field service?   Yes

(c) On duty?   Yes

(d) Off duty?   No

15. Was a Court of Inquiry held on the injury?

If so—(a) When?

(b) Where?

(c) Opinion?   NOT APPLICABLE

16. Was an operation performed? If so, what?   Yes

17. If not, was an operation advised and declined?   No

18. *In case of loss or decay of teeth.* Is the loss of teeth the result of wounds, injury or disease, directly* attributable to active service?   NOT APPLICABLE

19. Do you recommend

(a) Discharge as permanently unfit,   No
or
(b) Change to ~~England?~~ Australia   Yes

_____
Officer in medical charge of case.

I have satisfied myself of the general accuracy of this report, and concur therewith;
except†

_____ Major.
S.M.O. No. 2 AUSTN. COMMAND DEPOT.

Station _____
_____
Officer in charge of Hospital.

Date _____

No. 2 AUSTN. COMMAND DEPOT
A. M. C.
10 NOV 1916
WEYMCI
No.

* Loss of teeth on, or immediately after, active service, should be attributed thereto, unless there is evidence that it is due to some other cause.
† Delete this word if no exceptions are to be made.

Upper Berth.

Opinion of the Medical Board.

NOTES.—(i.) Clear and decisive answers to the following questions are to be carefully filled in by the Board, as, in the event of the man being invalided, it is essential that the Commissioners of Chelsea Hospital should be in possession of the most reliable information to enable them to decide upon the man's claim to pension.

(ii.) Expressions such as "may," "might," "probably," &c., should be avoided.

(iii.) The rates of pension vary directly according to whether the disability is attributed to (a) active service, (b) climate, or (c) ordinary military service. It is therefore essential when assigning the cause of the disability to differentiate between them (see Articles 1162 and 1165, Pay Warrant, 1913).

(iv.) In answering question 20 the Board should be careful to discriminate between disease resulting from military conditions and disease to which the soldier would have been equally liable in civil life.

(v.) A disability is to be regarded as due to climate when it is caused by military service abroad in climates where there is a special liability to contract the disease.

20, (a) State whether the disability is the result of (i.) active service, (ii.) climate, or (iii.) ordinary military service.

active Service

(b) If due to one of these causes, to what specific conditions do the Board attribute it?

G S wound.

21. Has the disability been aggravated by

(a) Intemperance? *No*

(b) Misconduct? *No*

(c) Any of the conditions mentioned in Question 20. and if so which? *No*

22. Is the disability permanent? *No*

23. If not permanent, what is its probable minimum duration?

To be stated in months.

six months

24. To what extent is his capacity for earning a full livelihood in the general labour market lessened at present?

In defining the extent of his inability to earn a livelihood, estimate it at ¼, ⅓, ½, or total incapacity.

½

25. If an operation was advised and declined, was the refusal unreasonable?

Not applicable

26. Do the Board recommend

(a) Discharge as permanently unfit, or

(b) Change to England?

Finding *Is for now unfit for general service for more than six months, is unfit for home service.*

Signatures

... D Lee ... President.

g macdonald Cpl RMC

Members.

Station

Date

Stamp: No. 2 AUSTN. COMMAND ... A.M.C. 10 NOV 1916 ... PLYMOUTH

Approved.

Station

Date

Stamp: DIRECTOR MEDICAL SERVICES 30 NOV 1916

... Chen

Administrative Medical Officer, *Major,* for Surgeon General.
D.M.S. A.I.F.

COMMONWEALTH OF AUSTRALIA.

W.B.

Department of Defence.

MELBOURNE. 7th Mar., 1917.

Dear Madam,

I am in receipt of cable advice to the effect that No. 4797 Private J. J. Griffin, 51st Battalion, is returning to Australia and is due in Fremantle about the end of March, 1917. It is regretted that the movements or name of the transport on which he is arriving cannot be disclosed.

It is to be noted that owing to possible mutilations in the cabled advice and other causes this notification may not be correct pending verification from the roll on arrival of the Troopship.

Yours faithfully,

J. M. LEAN. Major.

Officer i/c Base Records.

Mrs. E. Jones,
C/o Messrs E. Watkins Limited.
York Street,
ALBANY. W.A.

Transferred to

AUSTRALIAN IMPERIAL FORCE.

D

No. 4797

Rank P/c   Name GRIFFIN J.J.

Unit 51st (late 11th) Battn

Casualty Wounded GSW 26/9/6996 ae D. London 5 9 16 Ent
GS Wd (med) 15-9-16. Adm to one Eastern Gen Hosp WAMS London 14/9/06k 13-9-16
9.11.16 Discharged 1st Aus Dep to Weymouth H 80/33 2° London 4.11.16 Ey
Returning to Aust 4th HS Benalla left Plymouth 13.2.17 Due Melbne about 31.3.17 List
1070/8317 D° London 19.2.17

DATE.	PURPORT.	REF. NO.
4 SEP 1916	N.O.K. Advised Wounded	
SEP 29 1916	N.O.K. Advised in Hospital 16-9-16	
7 MAR 1917	N.O.K. Advised Returning to Australia	
13/3/17	COPY MADE FOR WAR PENSIONS	
	Returned 5 M.D. "Benalla" gro kid upper arm forearm	
	Discharged 5 M.D	

WAR HISTORY INDEX

1914/15
STAR
N.E.

BRITISH WAR MEDAL
TO
ISSUED
9416

VICTORY
MEDAL
ISSUED
No 9223

LIST.

Appendix II. People

Jack

John Joseph ("Jack") Griffin was born on 9th February 1881 in Co. Kerry, Ireland. He worked for a few years in an ironmongers in Killorglin, Ireland but left Ireland in December 1900, arriving in Australia in February 1901. Jack worked as a drover on his uncle's station at Maroonah, Western Australia and later bought the station from his uncle's widow. Jack married Mary Glass in 1909 but Mary died shortly after their marriage and they had no children. During World War I he served with the Australian armed forces and was injured on the battlefields of France in 1916. He died in Roebourne Hospital, Western Australia on 20th May 1926.

Parents

Jack's father, Edward Griffin, was born in Co. Cork, Ireland in 1834. He was baptised and brought up a Roman Catholic but later converted to the protestant faith. Edward was a school teacher in Killorglin, Co. Kerry. He married Jack's mother, Elizabeth Mansfield (1956-1902), on the 18th of April 1880 in the Church of Ireland parish of Killorglin. In later years Edward lived with his youngest son, also called Edward. Jack's father died in 1916.

Brothers

Jack had two brothers. George Griffin was born on the 13th November 1885 but died shortly afterwards. His other brother, Edward Blake ("Ned") Griffin was born on the 1st December 1886. Ned married Mary Jane Agnew and they had nine children. Their descendants live today in Ireland, the UK, Africa and Australia.

Sister

Jack's sister, Margaret Catherine ("Madge") Griffin was born on 2nd July 1882. She married Joseph Edward Naughton and had a son, Thomas Edward ("Tom") Naughton. After the death of her husband Madge emigrated to Australia with her young son and lived on Maroonah Station. In 1913 she married Edward Jones, a Methodist minister from England and they had three children.

Uncle

Born in 1844, John Harman Mansfield, the brother of Jack's mother, emigrated to Australia around 1872. Initially he worked as a drover and later became the owner of Maroonah Station in Western Australia. He married Annie McCafferty (1854-1911) but they did not have children. John died in 1907.

Cousins

Jack's cousin George Harman Dixon was educated at the Charter School, Kilkenny, Ireland along with Jack's brother Edward who was the same age (1901 census data). Originally from Enniscorthy in Ireland, George Dixon emigrated to Australia and worked with Jack on Maroonah Station. He served with the army in World War I and later obtained a soldier settlement farm in Western Australia, married a local girl and had three children.

Another first cousin, Anne Grey, from Booterstown in Dublin, also lived for a time on Maroonah Station.

Nephew

The son of Jack's sister Madge, Tom was born in Ireland but spent his childhood on Maroonah Station in Western Australia. He later worked for George Harman Dixon at his farm. At one time Tom owned a claim in the gold fields of Western Australia and mined asbestos in Yampi Gorge. He later bought a small dairy farm and married his wife Ellen Mary ("Marie") in 1941; they had a daughter, Joan. During World War II Tom served as a corporal.

Appendix III. Ships

Duke of Norfolk

The *SS Duke of Norfolk* was originally called the *Nairnshire*, and in later years was renamed *Marcellus*, then *Johanna* and finally *Pericles*. The *Duke of Norfolk* was the ship that carried Jack from his home in Ireland to Western Australia, a journey that lasted two months. He arrived in Australia in February 1901. The *Duke of Norfolk* was 3819 gross tons, 350 feet long and 48 feet wide, had a steam triple expansion engine, a single screw and a service speed 10 knots. She was built in 1889 by Hawthorn Leslie at the Hebburn Shipyard on Tyneside, UK. On the 24th May 1914, just two months before the outbreak of World War I, she foundered at the Western end of the English Channel after striking submerged wreckage.

Ulysses

In 1916 the *Ulysses* carried Jack and his battalion from Western Australia to the battlefields of Europe. Launched in 1913, the *Ulysses* was the largest ship to serve as a troop carrier during World War I. She weighed 14,499 tons with an average cruise speed of 14 knots. The *Ulysses* was owned by the China Mutual SN Co., London and leased by the Commonwealth until the 15th August 1917. She was used again to ferry Australian troops and airmen during the Second World War. The *Ulysses* was torpedoed by a German submarine in 1942 and sunk after apparently disobeying an order that would have led her through safer waters.

Stad Antwerpen

After Jack was wounded during the battle for Mouquet Farm, he was evacuated from France to England on the hospital ship

HS Stad Antwerpen on 21st August 1916. The *Stad Antwerpen* was a cross channel turbine steamer on the Ostend-Dover service commandeered by British forces for use as a hospital ship.

Benalla

HMAT Benalla was the troopship that brought Jack back to Australia in 1917 after his hospital treatment for wounds received at Mouquet Farm during the Battle of the Somme. The *Benalla* weighed 11,118 tons with an average cruise speed of 14 knots or 25.92 kph. She was owned by the P & O SN Co, London, and leased by the Commonwealth until 6th August 1917.

Appendix IV. Wars and Battles

Boer War (1899-1902)

The Boer War was fought from 11th October 1899 until 31st May 1902 in southern Africa between the Afrikaans-speaking Dutch settlers and the forces of the British Empire. It ended with a British victory and eventual incorporation of the territory into the Union of South Africa, at that time a dominion of the British Empire. The conflict is commonly referred to as the Boer War but is also known as the South African War or the Anglo-Boer War.

World War I (1914-1918)

World War I, also known as the Great War, was triggered by the assassination of Archduke Franz Ferdinand, the heir to the throne of Austria-Hungary, by a Serbian nationalist in the city of Sarajevo. This war lasted from the 28th July 1914 until the 11th November 1918 and ended in an Allied victory. By the end of the war four empires (German, Russian, Ottoman, and Austro-Hungarian) had ceased to exist and the map of central Europe had been completely redrawn. More than nine million combatants were killed during the conflict.

Battle of the Somme (1916)

The Battle of the Somme refers to a series of battles fought between the 1st July and the 13th November 1916 along the Somme Valley in France. Due to the huge number of casualties the word "Somme" has since become synonymous with battlefield slaughter. By the time the offensive was abandoned in November, the allied forces had managed to advance only twelve kilometres. The battle resulted in around 500,000 German casualties and 620,000 Allied casualties.

Battle of Pozières (1916)

The village of Pozières in northern France was the site of fierce fighting between 23rd July and 7th August 1916. This has become known as the Battle of Pozières and was part of the Battle of the Somme. The village was completely destroyed during the battle but has been rebuilt. The Australian forces suffered over 5,000 killed, wounded or taken prisoner.

Battle of Mouquet Farm (1916)

Mouquet Farm was located just north of Pozières in northern France. The Battle of Mouquet Farm, which began on 5th August 1916, was part of the Battle of the Somme and followed the Battle of Pozières. Thousands of Australian troops died over a period of several weeks while the farm was taken and abandoned a number of times. The farm was totally destroyed and the site of the original farm building is today clear land; the farm buildings were rebuilt south of the original location. Jack's service record states that he was wounded by a gunshot wound (or shell wound) to the right upper arm and forearm on the 14th August 1916 at Pozières.

World War II (1939-1945)

World War II was a global conflict involving most of the world's nations. The war began on 1st September 1939 with the invasion of Poland by Germany. There were two opposing military alliances: the Allies and the Axis. It was the most widespread war in history and was marked by the mass bombing of civilians, the Holocaust and the use of nuclear weapons. The war resulted in over 50 million fatalities.

Index

www.ingramcontent.com/pod-product-compliance
Lightning Source LLC
Chambersburg PA
CBHW072143020426
42334CB00018B/1866